Charles William King

The Natural History of Gems or Decorative Stones

Charles William King

The Natural History of Gems or Decorative Stones

ISBN/EAN: 9783337025137

Printed in Europe, USA, Canada, Australia, Japan

Cover: Foto ©berggeist007 / pixelio.de

More available books at **www.hansebooks.com**

THE NATURAL HISTORY

OF

GEMS

OR

DECORATIVE STONES.

By C. W. KING, M.A.,

FELLOW OF TRINITY COLLEGE, CAMBRIDGE ; AUTHOR OF 'ANTIQUE GEMS,' ETC.

"HOSPITA SACRA FERENS, NULLI MEMORATA PRIORUM."—*Manil.*

BELL & DALDY, YORK STREET, COVENT GARDEN.
CAMBRIDGE: DEIGHTON, BELL, & CO.
1867.

PREFACE.

———◦◦◦———

In the Preface to the ' Natural History of Precious Stones, &c.,' I have fully stated the principles which have guided me in altering the arrangement of my subject from that observed in the First Edition, and in recasting my materials (largely augmented during the interval) into two separate and independent parts. But, for the information of those who may read the present volume alone, and might otherwise be surprised at its apparent incompleteness, it is necessary to mention beforehand that *this*, omitting the precious, treats exclusively of the commoner species of decorative minerals—the term " Gems " being used for want of any more definite appellation, to designate such as have been claimed as its special dominion by the Glyptic Art of all ages in the world's history, while at the same time subserving, though in a less degree, the purposes of the jeweller. The previous Part, on the other hand, comprises the History of the rise and progress of Mineralogy, as far as regards this province of the science, from the earliest times down to the seventeenth century; the Description of Precious Stones properly so denominated; of the Precious Metals considered chiefly in their relation to art; and of the more important remains exemplifying this relation anywhere preserved. In both divisions the same method of treating the subject has been pursued; but the very

nature of the articles seemed of itself to suggest the propriety of classing them in two separate groups (especially
when the increase of matter necessitated two volumes for
their republication) instead of discussing them in mere
alphabetical sequence as I had done at first.

To avoid unnecessary repetition, reference is occasionally made in this volume to its companion on 'Precious
Stones,' but to no great extent, it having been my object
to render each division of my subject complete in itself.
For the same reason, some observations properly belonging
to the other Part, have been reintroduced in this, when
absolutely required for the elucidation of the points under
consideration.

It would be most ungrateful for me to close my labours
in this field of ancient and modern science without availing myself of the occasion to acknowledge great obligation
to Mr. Maskelyne, Professor of Mineralogy at Oxford, for
the ready kindness with which he has constantly favoured
this pursuit of mine; by giving his judgment upon the true
character of antique materials that I have from time to
time submitted to him in all cases of doubt; by the free
access he has allowed me, for the purpose of verifying
specimens, to the rich store of mineral treasures so admirably displayed in the department of the British Museum
over which he presides; and lastly, for numerous valuable
suggestions that have guided me in many difficult questions
to sound and satisfactory conclusions.

During the time these sheets were passing through the
press I fortunately obtained permission, for which I have
to return most sincere thanks to Mr. Newton, to minutely
examine, with exemption from all embarrassing restrictions,

the Blacas Gems recently added to the department of our National Collection under his care. The long-established celebrity, and the importance of this Cabinet, the public interest which it has so strongly, and to an unlooked for degree, excited since its happy acquisition by ourselves, and what is more to the present purpose, the skilful and judicious measures which have been taken to render these treasures really "publici juris," and to give all interested in ancient Glyptics the power of studying them with facility and pleasure, are considerations which have induced me to quote from it in illustration of my remarks as largely as the advanced stage of the printing would allow. So rich is the collection in examples of the finest quality of the productions on which I treat, that infinitely more information of the highest value might have been derived from its contents had they been accessible to me at an earlier period of my undertaking.

C. W. KING.

Trinity College,
  April 29th, 1867.

(G.)   b

CONTENTS.

	PAGE
ACHATES : AGATE	1
AËTITES : EAGLE-STONE	11
ALABANDICUS : GARNET	14
ALABASTRITES : ALABASTER	22
AMETHYSTUS : AMETHYST	27
ASTERIA : GIRASOL-SAPPHIRE	37
BASANITES : BASALT	40
BATRACHITES : TOAD-STONE	43
BERYLLUS : BERYL·	50
CALLAIS, CALLAINA : TURQUOIS	57
CAMAHUTUM : CAMEO	70
CERAUNIA : RUBY'; BELEMNITE	77
CHALCEDONIUS : CALCEDONY	82
CHRYSOCOLLA	88
CHRYSOPRASIUS : CHRYSOPRASE..	90
CHRYSOLITHUS : ORIENTAL TOPAZ	93
CORALLIUM : CORAL	100
CRYSTALLUS : ROCK-CRYSTAL	104
CYANUS : LAZULITE	123
EUMITHRES : AMAZON-STONE; MITHRAX ; LABRADOR	127
GAGATES : JET	130
HELIOTROPIUM : BLOODSTONE	133

PAGE

JASPIS: CALCEDONY OF VARIOUS COLOURS 137

LAPIS LYDIUS: TOUCHSTONE 152

LYNCURIUM: JACINTH; ESSONITE 160

MAGNES: LOADSTONE 169

MOLOCHITES: GREEN JASPER: MALACHITE 176

MURRHINA: AGATE-VASES 179

NAXIUM: EMERY 192

LAPIS NEPHRITICUS: JADE 203

OBSIDIANUM: OBSIDIAN 209

ONYX 213

OPALUS: OPAL 239

OSTRACIAS: PYRITES 248

OVUM ANGUINUM: ADDER-BEAD 251

PANTARBES: RUBY 259

PORPHYRITES: PORPHYRY 261

PRASIUS: PLASMA 265

SANDASTER: MATRIX OF OPAL 269

SAPPHIRUS: LAPIS-LAZULI 273

SARDIUS: ORIENTAL CARNELIAN 278

SARDONYX 287

SOLIS GEMMA: MOON-STONE 300

SUCCINUM: AMBER 303

TOPAZIUS: PERIDOT 312

ZMILAMPIS: CAT'S-EYE 319

VITRUM ANNULARE: PASTES 323

DESCRIPTION OF THE WOODCUTS 365

INDEX 371

NATURAL HISTORY

OF

ANCIENT AND MODERN GEMS.

ACHATES : ᾿Αχάτης : *Agate and Jasper.*

THEOPHRASTUS in his notice of the gems used for signets
(31) has : " A handsome stone, too, is the Achates, brought
from the river of that name (the Drillo) in Sicily, and is
sold at a high price." But Pliny observes (xxxvii. 54) :
" The Achates was anciently in high estimation; now in
none at all. Found, at first, in Sicily on the banks of a
river so called, but afterwards in a great number of other
places, exceeding in size, and fruitful in varieties."

Sicily, as Castellani informs me, continues to supply the
Neapolitan lapidaries with abundance of Agates and Jaspers of every kind. Assyria, likewise, furnished ·the
Greeks with Agates in the times of Dionysius Periegetes,
from the bed of the Choaspes, washed down from the
mountains by the winter-rains. Of the Malwa mines, the
most productive of all, a detailed account will be given
in its appropriate place under ONYX. But as regards
curiosity, the most astonishing workshop of Dame Nature
for the production of these stones, in endless and fantastic
variety, is to be seen in the desert valley, some seven miles
distant from Cairo, popularly believed the original bed of
the Nile. Besides these gems that lie about in the shape

(G)   B of water-worn pebbles, including the remarkable "Egyptian Pebble" (to be noticed hereafter), whole trunks of trees are to be observed retaining their natural characters, but completely metamorphosed into semi-Opal and Agate.

The stone intended by Theophrastus had evidently, from the terms in which he mentions it, no more than one sort; but, as he gives no definition of it, its character remains a matter of conjecture: certain it is that the modern "Agate" never presents itself amongst the Glyptic remains of his times. One thing, however, can be asserted positively, that it was *not* the stone now known as the "Tri-colour Agate," because formed of bands of light-brown, transparent-white, and opaque-black; each band sharply defined in strong contrast to its neighbour; that being his ὀνύχιον (ONYX). On such a stone, next to the Sard, occur all the best engravings of the early Greek and Etruscan schools; but hardly ever any works in the peculiar style of Imperial Rome.

"Agate," in the modern nomenclature, signifies an impure kind of Calcedony, presenting the most brilliant and varied colours, arranged in opaque wavy lines around a crystallised centre, and upon a *translucent* ground; thus being distinguished from Jasper, which, though much of the same nature, is always *opaque*, and contains a larger proportion of iron. No better definition can be found for it than that given by De Boot (ii. 95): "Next in appearance and in colour to the Onyx is the Agate. The Onyx is adorned with *zones* of colour; the Agate is *not*. For, instead of zones, it is ornamented by Nature in a wonderful manner with lines or spots of various colours, which exhibit images of different objects; some, for instance, represent, far from obscurely, trees, animals, fruits, flowers, clouds, &c." But the Achates of the Romans was a much more comprehensive term than that of the Greeks, or even of

modern mineralogists. Besides the species still retaining
the name, it embraced all the inferior quartz-gems, the
clouded Carnelians, and the like, as well as those now called
Jaspers : for the ancient name "Jaspis" was properly restricted to the green translucent species of Calcedony, the
Plasma, and the Heliotrope (JASPIS). How wide the signification of the term "Achates" had become in ancient
times appears from Orpheus (604), who sings its praises as
"the multiform, excellent Achates, invested with every
colour; for numerous to be seen are the dyes of the
Achates. In it thou wilt find, on looking, the glass-like
Jaspis, the blood-red Sard, and the sparkling Emerald.
Amongst them is found one of a vermilion aspect, moreover
a copper-colour, and the hue of the summer apple; but
chiefest of all that, canst thou obtain it, picturing the
tawny hide of the fierce lion dappled all over with spots;
red, white, dusky, and green,"—this last sort evidently
the Brocatella. These varieties are all distinguished in
Pliny by compound names expressing the different shades
or conformation of the subject-matter. Thus his Jasp-
achates would be an Agate with a green ground; his
Cerachates, one yellow and opaque; his Sardachates, one
mixed with translucent red; his Hæmachates, our rare
Red Agate, of the deepest blood-colour, traversed by
translucent veins of a lighter sanguine; the Dendrachates,
"decorated with little trees," our Moss-Agate.* The
Corallachates, "besprinkled with gold-dust like Lapis-
lazuli, and entitled *sacred*," is unmistakably our Aven-
turine-quartz, of a translucent reddish-brown, filled with

---

* Sometimes called "Mocha-stone," and consequently believed to
come from Arabia. But in reality "mocha" is merely the Saxon
patois for "moss;" the stone having the same name in both languages.
The filaments giving the appellation are not petrified vegetation as they
appear, but particles of clay suspended in the pure silicious matter of
the Agate whilst yet in a liquid state.

gold-like particles of Mica (SANDASTER). This, says Pliny, was found abundantly in Crete : at present the finest speci-mens are brought from Cape de Gatte, in Spain.

. That amongst these varieties our Red Jasper was included may be suspected from the singular pretence of the Magi "that the single * coloured sort rendered athletes invincible : the test being that, if boiled in a pot of oil along with other pigments, it should within two hours reduce the whole liquid to its own vermilion dye." That Pliny's Leucachates and Cerachates were whitish and yellowish Calcedonies, seems borne out by his remark, "jam e cerea staticula, et equorum ornamenta." The staticula, properly "steelyard-weights," were possibly those heads in Calce-dony, still very abundant, the perforations through which show the intention of attaching them to the dress, perhaps as weights to the ends of robes, in order to make them set well, like the metal ῥοῖσκοὶ terminating the Grecian chlamys. And the mention of "horse-ornaments" explains the existence of the larger specimens rudely worked into Isis or Gorgon-heads, and traversed by large diagonal holes evidently intended for a stout cord. Now a Pompeian bas-relief of a negro in a biga, led by a warrior, displays very conspicuously on the breast of each horse a grand Medusa's head in front-face, similarly suspended. This sculpture illustrates Pliny's words; and both, taken together, throw light, equally, upon the original purpose of these singular Glyptic remains.

The Agates from India were greatly admired as *lusus naturæ*, "because they represented rivers, groves of trees,†

* "Unius coloris," probably a corruption of "sanguinis ;" which seems implied in the *vermilion* tinge it produced.

† The Agates of Œta, Parnassus, &c., were "similes limitum floribus," a description out of which no sense can be extracted. "Limitum" must be a corruption of "illilæ," which would give us "painted with flowers," a definition consistent with nature and with Pliny's style.

cattle, and even more defined objects, such as chariots, pilasters, and the ornaments worn by horses." All these figures were suggested to the imagination by the capricious arrangement of the veins in many Agates, notably in the " Egyptian Pebble," which latter, indeed, often presents images so accurately defined that it is difficult to believe them the mere unaided freaks of Nature. In the Case of Jaspers, in the British Museum, may be seen the exact portrait of Chaucer in such a pebble; the Strawberry Hill Collection possessed another of Voltaire; De Boot had one no larger than the nail of the middle finger, marked with a perfect circle of a brown colour, within which was the veritable figure of a bishop with his mitre on. On turning the stone a little, a different image became visible; and again turning it, two more, of a man and a woman, &c. Numerous other examples equally illusive will occur to the recollection of every mineralogist.

Epiphanius, though so much later than the Roman naturalist, appears to have derived his notice of the Achates from some much earlier Greek source belonging to the times when the name was confined to one particular species. " Eighth Stone (in the Rationale) the Achates. This has been supposed to be that called Perileucos, already described under ' Hyacinthus.' This is an admirable gem, somewhat black (or blue) in colour, having externally a white zone, like marble or ivory, running round it. This, too, is found in Scythia. And amongst these there is an Achates having the colour of a lion's skin, which, powdered and mixed with water, smeared upon the bite of any reptile, counteracts the poison of the scorpion, the viper, and such like things." Isidorus also (Orig. xvi. 11) makes but one species of the Achates: " It is a black stone, having in the middle circles of white and black joined together and variegated; and likewise resembling the Hæmatites." And again it is more clearly defined by Marbodus as "a black stone girt by a zone of white;" which expresses, as it would seem, the first part of Epiphanius' article.*

One would conclude from Theophrastus' mentioning it in the next following paragraph, that he regarded as a novel variety of the Achates the stone found in the gold-mines at Lampsacus, which, on account of its beauty, was cut into a signet and sent for a present to the king (Alexander) at Tyre. The Macedonian hero was, as appears from his exclusive patronage of Pyrgoteles, a connoisseur in engraved gems, like Julius and Hadrian after him.

In the Roman times, after the stone had completely gone out of fashion for signets, it was in higher repute than ever, on the score of its medicinal and talismanic virtues. The fanciful patterns so often drawn upon its surface by the finger of Nature had suggested to the superstitious Magi, the authors of all such notions, the idea of some wondrous occult virtue residing in the substance thus conspicuously signalled out by the Creator. Hence Orpheus sings (230) how, "if thou wear a piece of the Tree-Agate upon thy hand, the immortal Gods shall be well pleased with thee. If the same be tied to the horns of thine oxen when ploughing, or about the ploughman's sturdy arm, wheat-crowned Ceres shall descend from heaven with full lap upon thy furrows." And again (604) how "every kind is an antidote to the asp's bite, if taken in wine; but the more potent Leontoseres, if merely tied over the wound, cures the scorpion's sting; enables the wearer also to gain

* This must have been also the nature of the sole ring-stone noticed by Pliny as the production of Italy, the *Veientana*, so called because found at Veii, "in which a white boundary divided a black ground." Similar Agates very likely are yet to be picked up in the pebbly bed of the brook Trebia, that yet murmurs around the precipitous sides of the plateau which formerly supported the ancient rival of Rome.

the love of women; obtain his petitions from princes; and heal the sick man whose thread has not actually been severed by Clotho." And Pliny jocosely adduces the Magi as teaching "that *in Persia*, by burning the lion's hide sort storms could be averted, and thunderbolts to boot;" the proof of its efficacy being that, if thrown into a caldron of boiling water, it immediately cooled the same, but in order to do good, the gem ought to be strung on the hair of a lion's mane. On the contrary the kind coloured like the hyæna's skin they viewed with detestation as the exciter of domestic strife. The Corallachates was an antidote against spider and scorpion stings, a virtue for which Pliny was quite ready to give full credit to the *Sicilian* species, because the air of that island was fatal instantaneously to those insects. The Tree-Agate was good for the sight, and was used by physicians for palettes whereupon to grind down their medicines.

To these virtues the mediæval sages added others more transcendental. The Agate besides bestowing health upon the wearer, gained him the favour of heaven, and of mankind. Marbodus, interpreting Virgil's "fidus Achates" literally, ascribes the escape of Æneas from all his perils to the potency of this talisman which he always carried about with him. Others again, with more learning and equal want of sense, derived the name from ἄχος, "because the gem protected the wearer against all pain and trouble."

Immeasurably the first amongst the "nature-pictures," instanced above, would stand the celebrated Agate of King Pyrrhus, if the description of it given by Pliny (xxxvii. 3), from tradition ("dicitur") were indeed to be received as true. It displayed Apollo holding his lyre, attended by the Muses Nine, each with her proper attribute! all exactly depicted by the native shades and veins of the substance, and without the slightest assistance from art. A suspicion irresistibly obtrudes itself that such a group, though actually existing in the royal ring, was nothing more than a cameo-engraving, but passed off by the jocose Greek, sporting with their simplicity, upon the Roman envoys, utter novices then in art, as an unparalleled miracle of Nature.

To give a few examples of similar deceptions in later times. The Shrine of St. Elizabeth (Marburgh), made in 1250, had for its chief glory, amongst the innumerable precious stones embellishing its surface, a large Cameo set above the statuette of the Virgin. It was a fine Sardonyx of three layers, representing the heads of Castor and Pollux, and was through the whole extent of the Middle Ages regarded as an inestimable production of *Nature*, and for which a former Elector of Mayence once offered in exchange, but without success, the whole village of Anemöneburgh. Even Camillo, the contemporary of Lorenzo dei Medici, is still so far under the influence of Gothic notions as to admit the *possibility* of the existence of similar lusus Naturæ. "In the second manner, stones seem as if engraved by the hand of Nature, when a portion of one stone adheres to the surface of another stone: or again, when some parts out of the same stone are deficient, through which addition or diminution a certain figure is produced, as is done by art in the case of *Camei* (in chamainis). And so, after this manner, it is possible for there to be stones engraved by Nature as well as engraved by Art" (III. p. 174.) Again, that Camei were popularly regarded in Gothic times as the work of sportive Nature is almost demonstrable from the way in which Agricola notices one then shown in the Church of the Three Kings, Cologne. He terms it an *Onyx*, more than a palm wide, in which the white veins mark out the heads of two youths, with a serpent in black, running from the one forehead to the other; also another Onyx with a negro's head, bearded, in black.

To return to the Agate of Pyrrhus, it is a singular coincidence that, when Camei first came into fashion in this country for ornaments, they went by the name of *Agate-stones.* Shakespere, for example, puts into Hero's mouth the simile—

" If low, an Agate very vilely cut."

Not to mention his hackneyed description of Queen Mab—

" In shape no bigger than an Agate-stone
On the fore-finger of an alderman."

In the latter passage the comparison as to *shape,* where one of *size* would be the more natural sequence, must necessarily refer to the *shape* cut upon the Agate, the tiny Cameo Nymph or Bacchante, those favourite subjects with the artists of the day. Amongst Queen Elizabeth's jewels is entered : " An *Agath* of her Majesty's visnomy." Similarly Vanderdoort calls his catalogue of Charles I.'s Collection, " an inventory of the pictures, coins, and *Agates,* &c."

Nevertheless, it will be owned by all mineralogists who have attended to this particular, that Nature has really sometimes drawn such artistic pictures upon the Agate as to predispose us not altogether to reject as a mere fable the account of that marvellous ring of the Epirote prince. In fact, an Agate was lately presented to myself, which contains within a natural circular frame a charming little sea-view, taken by moonlight. In the foreground lie craggy rocks, then comes the sea, receding gently in parallel lines with the due gradations of shade ; in the far distance rises a rocky island supporting a pharos ; the crescent-moon riding high in the heavens illumines the scene, aided by a twinkling star or two. The whole, painted in different shades of brown and white, is full of a dreamy repose, and of a feeling strongly recalling Bewick's inimitable little cuts in the same line. Camillo also mentions that the doorway of the " study " in Cesare Borgia's palace at Pesaro was adorned with plaques sawn from a column (antique) of a certain variegated stone, which exhibited so many different objects, that no one could believe it who had not seen the same.

The Chinese *manufacture* monster Agates exhibiting the loveliest colours very fancifully disposed, by staining thin slabs of alabaster by some process of their own : and these fictitious lusus Naturæ are often to be seen in this country treasured up as wonderful curiosities of mineralogy. But far more wonderful was the invention of the Florentine anatomist (the secret died with him) of petrifying human viscera into real Agates. In the Hospital of S. Spirito may be inspected still by the incredulous a table-top made up of hearts, lungs, livers, &c., thus agatised into one large slab—meet board for a banquet of vampyres.

How firm was the belief of our forefathers in the prophylactic virtues of the Agate, is remarkably illustrated by a jewel presented by Archbishop Parker to Queen Elizabeth (now in the possession of Mrs. Barwick Baker). It is a large oval Agate engraved by a contemporary hand with an intaglio of Vulcan in his forge, with Venus looking on, and is set for a pendant. The parchment accompanying the gift contains a long list of the properties of the stone, winding up with the dedicatory couplet, which evidently was considered by the learned donor to contain the happiest of conceits :—

> " Regni ἄχος Elizabetha gerit ; Matthæus achatem
> Cantuar ei donat fidus dum vivit Achates."

AETITES : *Eagle-Stone.*

THE lively fancy of the Greeks discovered in the forms of many natural objects resemblances to other organic forms, and, by a step not very logical, ascribed to them inherent virtues analogous to this interpretation. Of this system the *Aetites* is perhaps the most illustrative example, and the one also that the longest maintained its ancient reputation. Pliny describes four species of it (xxxvi. 21). The first was egg-shaped, white, and filled with a soft sweet-tasted clay: this was the female. The second, reddish-coloured externally, contained a stony substance, and passed for the male. The third was filled with a sweetish sand. The fourth, the Laeonian, had inside it a crystalline core, called the *Callimus.* The best kind were asserted to be found only in the nests of eagles—which could not breed without their aid ; hence their name. They, for this reason, were of the greatest benefit to women in labour ; a notion which even Dioscorides appears to endorse. The substance itself, as the specimens figured by De Boot ' plainly show, was one of those calcareous hollow concretions, sometimes white, sometimes tinged with iron, well known to geologists, and which appear to be accidental formations, not petrifactions of older organised bodies.

The Aetites must not, however, be confounded with another stone named in a fragment of Theophrastus ἡ λίθος τῶν τικτούσων, " the *gem* of parturient women ;" to which he also alludes (in section 5) by, " The most wonderful and

most important property is that of the stones that bring
forth others."* This was the *Pæonites*, described by Pliny
under that name, as reported to conceive at a certain time
and bring forth another; on that account very serviceable
to women in labour; found in Macedonia, and in appearance
like water frozen. It was therefore a *gem*, as the feminine
gender in the Greek denotes, not a mere rough *stone*, like
the Aetites. Marbodus says of it—

> "The mountains of the Macedonian bold
> Within their mines the Pæonites hold;
> Unknown the cause, with imitative throes   ˋ
> It heaves and all the pangs of childbirth knows.
> Hence pregnant women its protection bless
> In that last hour when travail's throes oppress."

This faculty of propagating its own species was afterwards attributed by mediæval belief to the most precious
of all, doubtless as being the most worthy of such a pri-
vilege. " Ruæus relates that a lady of Heveren, belonging
to a noble Luxemburgh family, possessed two diamonds,
heirlooms, which *frequently* brought forth others, so that
whoever inspected them at certain fixed intervals could
clearly discern them giving birth unto an offspring of the
same nature as their own."

Of the Aetites another most singular property is alluded
to by Dioscorides,—that of detecting theft. His meaning
is explained by the traditional practice of the Greek monks,
described by De Boot from his own experience. " All the
persons suspected of the theft being called together, the
monk kneads up in their presence flour and water, sprinkled
with the powder of this stone, repeating at the same time

---

* A statement that the sceptic Pliny would not accept for proved,
though endorsed by the authority of his all-powerful patron Mucianus
' ter Consul.' "Idem Theophrastus et Mucianus esse aliquos lapides
qui pariunt credunt."

a certain incantation, and, moulding the paste into balls as
big as eggs, gives three to each, with a little drink. The
guilty party either finds it impossible to swallow one
mouthful, or is ready to choke in the attempt." By an odd
coincidence the Hindoos still employ a similar ordeal, but
substitute dry rice for the paste. No doubt, where all are
believers, the mere effect of the guilty conscience on the
nerves will suffice to hinder the culprit from getting down
the enchanted morsel; but De Boot affects a pious horror
at the whole proceeding, and ascribes its efficacy to dia-
bolical agency.

He attributes all its *medicinal* virtues to its natural attractive power; to which was due its effect as an antidote against
poison, which it through this property extracted from the
system. On this, too, depended its more special use; for, if
tied round the female's arm, it kept up the fœtus until ma-
turity, but, if transferred to her thigh, it attracted it downwards, and thus ensured a speedy delivery. Such was its
established reputation for this object, that, at the time he
wrote, ladies used to pay from ten to twenty thalers for a
genuine Eagle-stone.

ALABANDICUS: *Almandine: Precious Garnet.*

"Garnet" Lessing conjectures to be an Italian corruption of "Garamanticus," an inferior kind of the Carbunculus according to Pliny's classification; but a much less far-fetched derivation presents itself, viz., that the common gem has borrowed its present name (Anglicised from Grénat; Granato) from the *Granaticus* specified by Marbodus as early as the 11th century. This was the Red Hyacinthus of the Romans, so called from the resemblance of its colour to the crimson juice of the pomegranate. For stones of the same colour were promiscuously classed under one head by the ignorance of the Middle Ages (unacquainted with even the ancient test of comparative hardness), whence has arisen that strange interchange of names between ancient and modern precious stones so perplexing to every mineralogist. But in the present instance the confusion is the more excusable, seeing that every variety of the Red Hyacinthus (Ruby) has an exact counterpart in colour amongst the various kinds of Garnets, and in many cases they can only be distinguished from each other through hardness, specific gravity, original crystallisation, and other properties not obvious to the eye, till lately the sole criterion. The Father of Mineralogy, Theophrastus, evidently is describing several very different stones under the head of Ἄνθραξ (18); for although his first kind, "brought from Carthage and Massilia, blood-red, but like a live coal when held against the sun, and of extreme value, so that a very

small one sold for 40 gold staters (40 guineas)," seems to
have been the true Ruby, yet that found near Miletus in
"polygonal pieces" must have been our Garnet, the primary
form of which is the rhombic dodecahedron. Similarly those
named by him as found in various parts of Greece, and as
being of little value (33), and which he distinguishes from
the precious ῎Ανθραξ by giving them the lowering name
᾿Ανθράκιον, must have been the Common Garnet. These
were, that found at Orchomenus, darker than the Chian
sort, and out of which mirrors were made ; the Trœzenian,
red clouded with white ; and the Corinthian, still fainter in
colour. The plane surface of a dark Garnet will reflect
objects with considerable distinctness ; the "mirrors" here
mentioned were, it is not impossible, "table" Garnets of
this kind :* (the only gems employed in the Merovingian
jewels of the 7th and 8th centuries are in fact Garnet tables
neatly inlaid, so as to form patterns in the fibulæ of copper
gilt) : Theophrastus thereupon remarks that all the kinds
found in Greece were common enough and of trifling value,
but that the better sort were rare, and brought only from
few places—Carthage, Massilia, Egypt, and Syene. Pliny
divides his Carbunculi into male and female, the former
of a brilliant, the latter of a duller lustre. In the males of
the Carthaginian kind, as it were a blazing star shone
within them, whereas the females diffused their entire
lustre externally. These Carthaginian stones were smaller
than those coming from India. It may be deduced from
these characters that the male Carbunculi were our Rubies,
the females our Garnets.

---

* This is almost made out from Pliny's description of the Hæphas-
titis (which Marbodus adds was found in the Corinthian Isthmus), as
" although of a fiery-red colour, possessing the nature of a mirror in
reflecting the images of objects.

The precious varieties, according to Pliny, were the Indian, the Garamantic or Carthaginian, the Ethiopic, and the Alabandine. The last were so called because, though found in the Orthorian rocks, they were worked up (perficiuntur) at Alabanda. This was the sort called by Theophrastus the Milesian, both places being in the same province, Caria. The Almandines of the moderns are the finest species of Garnet, of a beautiful crimson tinged with violet, and are brought from Siriam in Pegu, hence they are vulgularly called *Syrian* Garnets. These were the Amethystizontes of Pliny, then, as now, considered the best of the whole species, though it is probable he, or at least his Greek authorities, included the *Balais* under the same designation. He notices how some amongst the males possessed a more liquid, others a darker fire ; how some were lighted up with a colour not their own, and shone more than others in the sunlight. The description he quotes from Archelaus of the Carthaginian sort exactly applies to our best Indian Garnet, " that it was of a darkish aspect, but kindled up more brightly than the others when held and turned about in the light of the fire or sun. In the shade within doors the colour was purple ; in the open air flamy ; the wax sealed with such a gem would melt even though in the cool."

These descriptions of the several varieties of the ancient Carbunculus will be rendered more intelligible by a brief account of the various Garnets known to the modern lapidary. First in value comes a rarer sort, the *Almandine*, already described. It is a splendid stone, of great lustre, and, when of the first quality, can with difficulty be distinguished from the purple Spinel, which indeed usually passes under the same name in its antique specimens. Fine Roman intagli frequently, and sometimes imperial

portraits, occur in this admirable material.* Closely approximating to the Jacinth in colour, a pure, pale orange,
and therefore usually confounded with it, is the *Essonite*, or
Cinnamon-stone, although in its composition a true Garnet.
It comes to our jewellers exclusively from Ceylon, but the
Vesuvian-garnet, found in small crystals in the cavities of
the lava flowing from that mountain, is exactly identical
with it, both in appearance and constitution. The *Common
Garnet* is precisely of the colour of Burgundy wine, more or
less diluted, according to its goodness. The *Pyrope* differs
slightly from this chemically (magnesia replacing in it the
peroxide of iron), but not in appearance, except in being
clearer and brighter. The *Carbuncle* is somewhat darker
than the last, and is, in fact, the common Garnet cut " en
cabochon," or " tallow-drop," to use the genuine old English
term, that is, into a very convex form on the upper surface,
whilst the base is hollowed out more or less to give translucency to the stones; for in their native state they are so
dark in tint as to be nearly black until held against the
light, when the red becomes visible. Many antique Carbuncles are found with the back hollowed out, precisely in
the modern manner; but, if the quality of the stone allowed
it, the ancients preferred cutting the under side of the gem
to a plane surface, instead of increasing the risk of fracture
to so brittle a material by diminishing its substance. The
Vermilion Garnet, or *Vermeille*, so called from its unmixed
red tint, is only a jeweller's name for the Pyrope, the Bohemian Garnet.

Garnets seem to have been little employed by the Greeks
for engraving upon, but were largely in favour with the
Romans of the Empire, though not at a very early date, as
may be concluded from the frequent occurrence of splendid

---

* A still rarer shade of the Almandine, but also known to the ancients, is of a pale rose-colour, much resembling the Balais, now exactly
imitated by the burnt Brazilian Topaz.

(G)   C

specimens completely disfigured by the wretched abortions
in the way of intagli cut upon them, evidently the productions of the very decrepitude of the art. Nevertheless, many
tolerable, and a few excellent intagli *do* occur on Garnet,
but for the most part on the Almandine, a testimony to
the superior estimation in which that variety has ever
been held. That very intaglio to which, as Köhler justly
observes, neither ancient nor modern art has ever produced
an equal as regards the skill and industry displayed in its
execution, the " Head of the Dog Sirius " (Marlborough), is
engraved in a perfect Indian Garnet of unusual size and
beauty. The impression from this intaglio presents the
head in full relief, with open jaws, the interior of the mouth
represented with miraculous fidelity; and its value still
further enhanced by the legend on the collar, ΓΑΙΟΣ
ΕΠΟΙΕΙ. The antiquity of this work has been disputed,
without much cause : certain it is that the artist Natter, to
whom it has been assigned, was far from capable of producing such a masterpiece. A smaller replica of the same
head on an Almandine is in my possession; the hot and
fiery nature of the stone was doubtless regarded as
analogous to the subject upon it, the blazing Dog-Star.
Another famous Almandine is that of the same cabinet,
engraved in the highest style of Roman art with the heads
of Socrates and Plato ; a gem which above all other antique
remains has served to identify the portrait of the latter
philosopher. A few fine heads besides in this stone might
be quoted, but such are of excessive rarity, and belong all
to Imperial times.

The Sassanian kings, however, frequently appear upon
this gem ; in fact, it would seem to be have been regarded
by the later Persians as a royal stone, from the preference
they have given it for the bearer of the sovereign's image
and superscription.

Callistratus states that some of the Indian Carbunculi attained such extraordinary dimensions as to admit of being formed into cups holding a sextarius, or nearly a pint. Such stones, according to Satyrus, were never clear, but generally foul within, and of a harsh, disagreeable tint. Such a description applies to our commoner Carbuncle, which does indeed attain a very great bulk, though the fact mentioned by Callistratus is tinged by the usual exaggeration of the Grecian traveller in the far East. Yet I have seen an antique cup, hollowed out of a solid Garnet, as large as a half goose-egg, and which was engraved internally with the name of its ancient owner, Codrus; * and a heart-shaped tablet (Hertz Collection), covered with a long Gnostic formula on both sides, an Alexandrian work of the 3rd century, is $2\frac{1}{2} \times 1\frac{1}{2}$ inches in extent, showing the immense magnitude of the crystal from which it has been sawn. In these two works the material was of bad quality, confirming Pliny's quotation from Archelaus (30), that "those found near Thebes were brittle, full of veins, and like burning coals nearly expiring." † These garnet bowls, or at least paste imitations of them, are certainly what Martial means by his "Amethystinos trientes" (x. 49); now, as the triens, according to its name, held four ounces of liquid, such bowls would have no more than equalled a small tea-cup in capacity. On the

* The inventory of the French Crown Jewels, drawn up in 1791, mentions, "An oval cup of a single Garnet, rich in colour, $3 \times 2\frac{1}{2}$ inches wide, and 3 high, valued at 12,000 francs; and a round cup of Oriental Garnet, full of flaws, 3 inches wide by $1\frac{1}{4}$ deep; a third is "vermeille d'Allemagne," $2\frac{1}{2}$ deep, at 6000 francs; besides six others of less size.

† To give a notion of the magnitude attained by this species it will suffice to adduce two specimens from the cabinet of the Marquis de Drée; one, a Siriam Garnet of the finest quality, octagonal in shape, was $7\frac{1}{2} \times 6\frac{3}{4}$ inches in measurement, and sold for 3550 francs. The other of a flame-colour, $10\frac{3}{4} \times 6\frac{1}{2}$, for 1003 francs (122*l.* and 40*l.*).

other hand, specimens of all kinds show that the usual
scyphus in silver, or earthenware, or glass, was quadruple
that size.

. Pliny concludes his long dissertation (of six chapters)
upon the Carbunculus with the observation that all the
varieties of this species obstinately resist the engraver, and
the wax adheres to them in sealing. This is quite correct
as regards the soft sealing material used by the ancients,
a composition like our modelling-wax, made of bee's-wax,
rendered plastic by the addition of a few drops of turpentine to the melted mass, and a little vermilion to colour it.

Although the great artists of the Renaissance have left
us a few magnificent works in Carbuncle, both in relief
and intaglio, yet modern artists have seldom employed it,
except for the small cameo portraits intended to pass,
when foiled, for done in Ruby. The stone is extremely
hard to engrave, and, besides this, very brittle, difficulties
that they cannot overcome; a thing which bears so much
the stronger testimony to the skill of the ancient masters,
who have left such highly finished works in so refractory
a material.

Our Garnets and Carbuncles, aluminous silicates coloured
by iron oxide, are now supplied in large quantities from the
mines of Zöblitz in Silesia, from the Tyrol, and from
Hungary. These are all inferior, both in beauty and
hardness, to those from Siam, which continues to send us
Garnets of the richest red, tinged with yellow, besides its
peculiar purple-tinted Almandines. The latter also come
in abundance from Ceylon; which gives them their popular
name of Ceylon-rubies, by means of which our jewellers
obtain a better price for them from the ignorant. In
spite of this abundance, even now a stone of a certain
size, of a fine rich tint, and free from flaws, is of considerable value, ranging from 8*l.* to 10*l.* One of five

carats is put down at 1200 francs in the inventory above quoted. But its estimation has greatly fallen since the days of Mary Queen of Scots, the pendent carbuncle to her necklace worn at her marriage with the Dauphin being worth 500 crowns—an enormous sum in that age. To the same effect De Boot, about the year 1600, states (II. 36) that small Bohemian Garnets, up to the bigness of a pea, were found abundantly in the fields around Prague, but that one the size of a hazel-nut would equal a Ruby in value. This kind he prefers to the Indian. The latter he estimates at 2 thalers* per carat up to 20 carats; then at 3 up to 40; at 4 up to 60; and at 5 up to 100. But in De Laet's time, fifty years later, both Indian and Bohemian were become of little value.

* The thaler of this date equalled our crown-piece in intrinsic, and at least its quadruple in current value in Germany.

ALABASTRITES.

This is the stone now known as the Oriental Alabaster (Carbonate of Lime), as is manifest from Pliny's description (xxxvi. 12) of the best sort, the Carmanian, "of a honey-colour, variegated with spiral spots (vortices) and not transparent; for the colour of horn, or fatty white, or anything resembling glass in it, are considered blemishes." To this the name of *Onyx* was originally given from the resemblance of its layers and tints to the shades in the finger-nail of a "well-bred person," to quote Epiphanius. This last writer mentions a conjecture of some, that its formation was due to the dropping of water: in which they were altogether in the right, for it is identical in constitution with Stalagmite. But this same ignorant transcriber concludes his article on the Onyx with a hopeless confusion between the *marble* and the *gem* of the same name. To avoid such ambiguity, the Romans finally restricted the term Onyx to the gem so called at present: one kind of which, the Agate-onyx (made up of layers of opaque and transparent white), exactly resembles that regularly stratified variety of this marble, the Albâtre-onyx of the French. The Onyx-marble now lost its ancient title, and became the *Alabastrites*, from its being chiefly employed as the best material for the Alabastra, or perfume jars, shaped like minute amphoræ, but "without handles," as their Greek appellation signifies. Such alabastra were made in all materials—pottery, glass, the precious metals; but this stone was above all the most in use. Hence St. Mark's ἀλάβαστρον μύρου νάρδου πιστικῆς,* and Horace's "Nardi parvus Onyx," meant one and the same thing, the latter retaining the ancient designation of the substance. The slender necks of these jars were readily broken off to come at their contents (perfumed oils), having been closely sealed down by the maker on leaving his laboratory. To their reputation for preserving the perfume unimpaired for a great length of time being perfectly deserved, a convincing testimony is offered by certain large alabastra from Pompeii (now in the Museo Borbonico) still diffusing a strong odour of their ancient contents: whereat the Emperor Nicholas on his visit "rimase sorpreso," as the custode tells you; and not without reason. The inferior Onyx-marble chiefly used for this purpose was quarried, says Pliny, near Thebes in Egypt, and Damascus. Vast numbers of Canopi, or sacred jars, of a squat form, with the head of a mummy for a lid, still exist in this Egyptian stone, which is identical in quality with the Derbyshire Alabaster so much worked up now into cheap ornaments for the plebeian mantle-piece.

This common Alabaster certainly deserves the name of "Finger-nail stone" better than the more precious substance, the stratified Agate, that has usurped its original title, for it often exhibits layers, slightly curved, of flesh colour and opaque white arranged like the shades in the human

---

* His "two-hundred denarii" was the regular price for the pound-weight of unguentum. Martial (xii. 65) debates with himself whether he shall reward a kind nymph with

,   "Utrumne Cosmi, Nicerotis an libram,"

or with
            "De moneta Cæsaris decem flavos,"

ten "yellowboys," that is with 250 denarii in value.

original. The Greeks however made a subtle distinction in the appellations of the two species, not observed by the Latin mineralogists; giving the name of 'Ονύχιον to the gem, and of 'Ονυχίτης to the marble. 'Ονυχιτὶς λίθος, in the feminine, to distinguish its precious quality, is used by Appian for the Murrhine-agate, in speaking of the treasures of Mithridates. (See MURRHINA).

The Oriental Alabaster, when first imported into Rome, was considered a highly valuable material, in fact almost a precious stone, thought to be produced in Arabia alone. Brought over in such small pieces as only to be worked up into cups and the feet of couches and chairs (the lion's-paws so often seen serving in that capacity), some amphoræ carved in it, as large as the Chian wine-vessels, exhibited by P. Lentulus Spinther, were regarded, says Corn. Nepos, as wonderful curiosities.* Yet, five years later, so greatly had the importation of it increased, he had seen actual columns 32 feet long of that very marble. Four middling-sized columns were placed by Balbus in his theatre as an unprecedented ornament; this was under Augustus, whereas Pliny's contemporaries had with him admired thirty such, of larger proportions, in the banqueting-hall built by Callistus, the most powerful of the freedmen of the Emperor Claudius. Superb examples of such magnificence are yet preserved in some of the older Roman churches, relics of the times alluded to by Pliny; but none have ever approached to the magnitude of those presented by Mohammed Ali to the new fabric of S. Paolo fuori le Mura, columns and pilasters, 40 feet long, each of a single block and the most beautiful quality. Under this Pacha the ancient Egyptian quarries had been re-opened, and furnished the material of which

---

* A most elegant example of this pattern may be seen in the Temple Room (Brit. Mus.), and also a cinerary urn in the same beautiful substance.

his sumptuous mausoleum at Cairo is exclusively con-
structed : a piece of extravagance beyond the ambition of
even Nero.*

It is singular that Pliny should have noticed no other
quarries of this marble as then worked besides those in
Asia and Egypt, for the mountain on which Volterra is
built yields large blocks of a variety richer in point of
colour than the Oriental, a warm brown variegated with
lemon vortices: his omission is the more unaccountable
as the stone had been much in use with the Etruscans, as
their sepulchral monuments remain to attest. The magnificent and huge vases, now exported from Tuscany,
belong to the Volterran fabrique. It is indeed true that
the Volterra stone differs chemically from the Oriental,
being a Sulphate of Lime or Compact Gypsum ; but such
an analytical distinction would have had no significance
in ancient mineralogy. Thus the *Lygdinus* subsequently
mentioned as dug up in Paros, but in small slabs, never
exceeding the measure required for a dish or a bowl, and
equal to the Egyptian stone as a preservative of perfumes,
appears to have been Compact Gypsum, from its distinctive
character adduced of superlative whiteness, "candoris
eximii." The fact that some of the sorts were burnt for
lime and used in plasters (as depilatories?) also indicates
that the Onyx-marble included both the Carbonate and the
Sulphate of Lime.

The French clearly distinguish the two species, con-

* Nero actually built the temple of Fortuna Seia enclosed within
the precincts of his "Golden House" out of a translucent marble, the
Phengites, then just discovered in Carmania. This edifice, when the
door was closed, was perfectly light within, though without a window.
In certain very ancient Italian churches (as S. Miniato) the windows
are filled in with single slabs of a translucent stone, which is not Talc,
and thus, probably, preserve an example of this Phengites.

founded under the common name *Alabaster* in English,
designating the hard Carbonate of Lime Albâtre Oriental
when dappled, and Albâtre-onyx when regularly stratified :
the softer Sulphate of Lime on the contrary (the common
European kind) is their Albâtre Gypseux, or Alabastrite.
The Albâtre-onyx has been frequently used by modern
artists for camei instead of the *pietra dura*, the Agate-onyx,
which it closely resembles in stratification, colour, and
texture.

AMETHYSTUS : 'Αμέθυσος : 'Αμέθυστος : *Amethyst.*

OUR Common Amethyst, and the stone (perhaps) generally designated amongst the ancients by this name, is nothing more than rock crystal coloured purple by manganese and iron, and on this account is more properly termed in modern mineralogy Amethystine Quartz. It is therefore of an entirely distinct species from the true Oriental Amethyst, a most rare and valuable variety of the Precious Corundum, and which is in fact a purple Sapphire, but its purple shows little of the red (*ponceau*) seen in the common Amethyst, being rather an extremely deep shade of violet.* The name of "Oriental" is, however, improperly applied by the English lapidaries to the Amethystine Quartz when very brilliant and of two shades of colour (qualities dis· tinguishing the Indian from the German), the true gem of the name, from its rarity, being known to very few among them.

The name "Amethyst," though most probably a mere corruption of the Eastern name for the stone, a trace of which seems preserved in the Hebrew Achlamath,† was by the fanciful Greeks interpreted as though formed in their own language, from ἀ μεθὺ "wineless," and on the strength

* The common Amethyst, formerly brought from Carthagena in Spain, and now only to be met with in old-fashioned pieces of jewelry, alone of its species exhibits this pure violet colour.

† Perhaps the true origin is, as Von Hammer suggests, the Persian "Shemest."

of this etymology the gem was invested by them with the
virtue of acting as an antidote to the effects of wine.*
Hence the point of several epigrams in the Anthology, as
that of Antipater's (or Asclepiades) on the signet of Cleopatra, an Amethyst engraved with the figure of Μέθη, the
genius of intoxication (ix. 752)—

> Εἰμὶ Μέθη, τὸ δὲ γλύμμα σοφῆς χερὸς ἐν δ' ἀμεθύστῳ
> γέγλυμμαι· τέχνης, δ' ἡ λίθος ἀλλοτρίη·
> ἀλλὰ Κλεοπάτρας ἱερὸν κτέαρ· ἐν γᾶο ἀνάσσης
> χειρὶ θεὸν νήφειν καὶ μεθύουσαν ἔδει.

> "A Mænad wild, on amethyst I stand,
> The engraving truly of a skilful hand;
> A subject foreign to the sober stone,
> But Cleopatra claims it for her own;
> And hallow'd by her touch, the nymph so free
> Must quit her drunken mood, and sober be."

Another, more briefly playing on the same fancy (ix.
748)—

> Ἀ λίθος ἐστ' ἀμέθυστος ἐγὼ δ' ὁ πότας Διόνυσος,
> ἢ νήφειν πείσει μ', ἢ μαθέτω μεθύειν.

> On wineless gem, I, toper Bacchus, reign;
> Learn, stone, to drink, or teach me to abstain!

Or, as Pliny explains the import of the name (xxxvii. 40),
" because these gems never come up to the colour of wine,
since before they touch it their lustre falls off into the
colour of the viola " (i.e. pink cyclamen).

Pliny divides the Amethystus into five kinds, the Indian
holding the first rank; others coming from Arabia Petræa,
Armenia Minor, Egypt, and Galatia; inferior sorts from
Thasos and Cyprus. The Indian displayed the precise
colour of the imperial purple; a variety of these " degene-
rated into that of the Hyacinthus (Sapphire), and was

---

* Mohammed Ben Mansur affirms that wine drunk out of an
amethyst cup does not intoxicate.

called by the Indians Sacondion, Sacon being their term for that particular tint; if still lighter, it took the name of Sapenos." The fourth sort was of a wine (we should say, "Burgundy") colour; the fifth and worst of all was so pale as to resemble Crystal. The most admired tint was where a slightly rosy hue shone out from amidst the purple, and became more conspicuous when viewed by transmitted light (in suspectu); such were distinguished by the title of Pæderotes (Cupids), or the "Gems of Venus," on account of the pre-eminence of their kind and their beauty.

The deeper the tint the less brilliant is the stone, for which reason the ancient engravers preferred the light-coloured variety, which of all gems, next to the Jacinth, possesses the greatest degree of lustre; or they may have used it for cutting upon on account of its greater cheapness, remarked above. That Amethysts [*] of a fine colour (now so worthless) were deemed too valuable by the ancients to have their substance diminished by the sinking of intagli into them, appears from many examples extant. They were either worn as mere ornamental jewels uncut, or else polished to an extremely convex form, presenting in their exact centre a diminutive intaglio, a Gorgon's Head, or a mask, in the nature of a talisman that augmented the supposed virtue without detracting much from the native beauty of the gem. Pliny notes the suitableness of all the Amethyst family for engraving upon (scalpturis faciles), a sufficient proof that no species of this stone was the Hya-cinthus (the common explanation of archæologists from De Boot to K. O. Müller), which Solinus with justice calls

---

[*] The rich Indian Amethyst evidently was then equally precious with the Sapphire ; Pliny undeniably regarded the latter as merely a variety of it; for this reason the two are often found set side by side in ancient jewels.

the hardest of all gems, and only to be touched by the Diamond-point.

Intagli of all dates and in every style occur upon Amethysts, but so much more generally on the pale sort that an engraving upon one of a rich dark colour may, on that very ground, be suspected as modern. Besides the foregoing remarks as to the high value of such a shade in antiquity, the modern artists have usually employed the Hungarian Amethyst as being now the most abundant, of which the tint is a fine reddish purple, though the lustre is far below that of the Indian. Amongst the few exceptions to this rule that have come under my own notice is a head styled of Mithridates, but, in my judgment of some Bactrian king of the line of Euthydemus; perhaps the noblest Greek portrait in existence, cut in a large Amethyst of the deepest violet colour, found a century ago in India; which, however, being doubtless the royal signet, rather corroborates than weakens the previous statement. For such a use the most precious material procurable would naturally have been selected; "ut alibi ars, alibi materia esset in pretio," to use Pliny's expression. Another fine Greek intaglio, a head of Pan, in front face, on a similar stone, the antiquity of which could not be called in question, was in the Uzielli Cabinet; and, above all, stands the unrivalled Marlborough Omphale, the first amongst the numerous repetitions of that favourite subject, where the Amethyst (of the Indian kind) possesses equal lustre and richness of colour. In a large circular, convex stone of this sort is engraved the Berlin Atalanta, justly styled by Winckelmann "une des gravures les plus parfaites qu'on puisse voir." The swift-foot nymph is figured running at full speed and holding down with both hands the folds of her voluminous peplum distended by the agitated air. Through its gauzy material the elegant contour of her whole body is distinctly visible. She is turning round her neck and looking back as if about to stop to pick up the golden apple thrown down by her competitor.

Heads, and even busts, both in full and in half-relief, often occur of antique workmanship in this stone: as some perfectly preserved remains show they served to complete statuettes in the precious metals. The grandest of Medusa-heads, the Blacas, is carved out of an Amethyst of the darkest violet, two inches in diameter. Although the Amethyst came into use amongst the earliest materials used by the gem-engraver, for we find in it an abundance of Egyptian charms (pendants for necklaces), in the form of vases, shells, hands, &c., and sometimes scarabæi, the last of Etruscan work also, and Roman intagli in it are sufficiently numerous, yet it is a singular fact that we rarely meet with works in the highest style executed in this material. Probably the superior kind was too precious to be so employed, whilst the paleness of the other and cheaper sorts was repugnant to the taste of first-rate artists.

But besides the stone known at present by the same name, there can be little doubt the Roman "Amethystus" included amongst its varieties a totally distinct species of gem—some kinds of our common garnets. This conjecture is supported by the authority of De Boot, who says (ii. 30), "Amethystus veterum nunc Granati nomen obtinet."*

---

* There can be no question that the Amethyst (purple quartz) of the moderns was considered by the ancients as one species of their Jaspis. In no other way can Pliny be understood (37), where, describing the latter, he thus classifies them : " Minus refert nationes quam bonitates distinguere. Optima quæ purpuræ aliquid habet; secunda quæ rosæ ; tertia quæ smaragdi." And such a classification is perfectly accurate, all being equally quartz crystals, variously coloured by different metallic oxides. Again, he alludes to another *Jaspis* resembling the Sard;

We cannot resist such an inference if we carefully examine some of the characters given by Pliny of certain varieties of this gem. Thus he describes the Indian as "having the exact tint of the royal purple, and the dyers direct their endeavours to produce the colour, taking this gem for their pattern. For it diffuses a hue softly gentle to the sight, neither does it flash upon the eye like the Carbunculus." Be it remembered that he has already described the best purple as the colour of clotted blood, dark in one aspect, bright red if viewed against the light.* Again, we find "the fourth sort has the colour of wine;" now Italian wine generally (and more especially that grown about Rome) shows the richest Burgundy colour, than which nothing more accurately expresses the deep hue of the common Pyrope. It is a manifest absurdity to suppose a comparison between the bluish red of our Amethyst and the unmixed red of various shades peculiar to any sort of wine.† Again, his "Amethystus" was exactly counterfeited by staining amber with either alkanet-root, or murex-blood; both *reds* with no tinge of *blue*. The Carbunculus of Pliny was doubtless our Spinel Ruby, and to the eye alone (the sole criterion of the ancient lapidary) the Oriental (Siriam) Garnet and the Spinel are almost undistinguishable from

another mimics the viola (or pink cyclamen); the Cappadocian was a sky-blue mixed with purple (ex purpura cœrulea), but dull and not lustrous. The last definition applies exactly to our German Amethyst.

* A description closely applying to what is now called the Carbuncle. This dark shade of the ancient Tyrian dye is well exemplified by the remark of Augustus, preserved by Macrobius amongst his other facetiæ. When that prince was finding fault with the darkness of some purple purchased to his order, and the vendor repeatedly bade him "hold it up higher and look at it," he retorted, "Must I then be always walking on a balcony if I wish the Romans to know I am richly attired?"

† Theophrastus includes (30) τὸ ἀμέθυσον in his list of ring-stones, adding that it is like wine (οἰνωπὸν) in colour.

each other, as I have had frequent occasion to observe in looking over examples of both species which have come down to us from Roman times with engravings upon them. Again, in no other manner is it possible to understand what kind of precious stone Heliodorus is describing as set in the king of Ethiopia's ring (Æth. v. 13). "And so saying, he put into his hands a ring, one of the royal jewels, an extraordinary and astonishing thing, the shank being formed of electrum, and the beasil flaming with an Ethiopian Amethyst, in size about the circumference of a maiden's eye, but in beauty far surpassing either the Iberian or the British sort. For the latter blushes with a feeble hue, and is like a rose just unfolding its leaves from out of the bud, and beginning to be tinged with red by the sunbeams. But in the Ethiopian Amethyst, out of its depth flames forth, like a torch, a pure and as it were a Spring-like beauty; and if you turn the stone about as you hold it, it shoots out a golden lustre, not dazzling the sight by its fierceness, but resplendent with cheerfulness. Moreover, a more genuine nature is inherent in the species than is possessed by any brought from the West, for it does not belie its appellation, but proves in reality to the wearer an antidote against intoxication, preserving him sober in the midst of drinking-bouts." This gem was engraved with a youthful shepherd and his flock, of which the tasteful bishop proceeds to give a pretty description, in which he again dwells upon the "golden" tints commingled with the flamy hues of the stone. The same interpretation must be put upon the more obscure language of the somewhat later Epiphanius in his 'Treatise on the XII. Stones of the Rationale,' where under IX. he gives "The stone Amethyst: this in proportion to its circumference (" excellence " must be the true reading) is of a deep flame colour, or sometimes paler, sending forth

(G)   D

out of the inside a vinous appearance. Of it there are
various species: one of the sorts is very similar to a clear
hyacinthus (Sapphire), the other to the murex-blood, *i.e.*
Tyrian dye. They are found in the mountains and on the
coast of Libya." Again, Pliny's definition of the *Pæderos*
suits no other gem so exactly as a particular variety of the
Almandine sometimes met with amongst antique Garnets,
the tint of which is truly roseate, not purple. One such
in my possession is engraved with a Cupid proudly bearing
off the spoils of the vanquished Hercules; there is good
reason to suspect that the popular name of the gem had
influenced the choice of the subject.

Throughout the Gothic period the common Amethyst
held the same rank as it had enjoyed with the Roman
jeweller, and continues to keep company with the Sapphire
in the ornaments of the priest or prince. Even as late as
the year 1600 a perfect Indian Amethyst is valued at half
the price of the Sapphire, viz., one thaler or crown for the
first carat. For higher weights De Boot gives a curious
rule; to add together the weight, and value in thalers, of
the stone preceding in his table the number of the one in
question, and thus brings up the value of one of 20 carats
to the high figure of 201 thalers.

Even in the last century this now despised stone was
held in high estimation, when Queen Charlotte's necklace
of well-matched Amethysts, the most perfect ever got
together, was valued at 2000*l.*; at present it would not
command as many shillings, so swamping has been the importation of late years of German Amethysts and Topazes
(purple and yellow crystals of quartz), which are got in
endless abundance from various parts of Hungary, Bohemia,
Saxony, and notably at Oberstein, where they are cut and
polished expeditiously and cheaply by water power (on
sandstone wheels turned by the stream of the Nahe), and

despatched into all parts of Europe to be made up into cheap articles of jewelry. They are also found plentifully about Wicklow, in Ireland. Barbot mentions a crystal of Amethyst as recently brought to Paris of the astonishing weight of 65 kilos. (about 140 lbs.). When the gem was in fashion, it was formerly imported largely from the East Indies, and these were light coloured, the purple being shaded unequally, but extremely lustrous. The·colour of the Amethyst can be dispelled by a careful roasting in hot ashes. Hence in the last century, when it was the great desideratum with the jewellers to obtain a suite of stones all exactly of the same tint, they were able to bring about this result by subjecting the several pieces to the heat for a greater or less time, until they were all reduced to the same shade of purple. According to modern usage this is the only gem it is allowable to wear in mourning.

The artists of the Renaissance eagerly availed themselves of these huge and beautiful crystals to carve them into those fanciful yet elegant vases, so acceptable to the taste of their age. The Parisian Collections offer the choicest specimens of their skill in this line. A cup, shaped as a shell, seven inches long and deep, by six wide, valued at 1800 francs; also an urn eight inches high, fluted and elaborately decorated with engravings, are enumerated in the former treasury of the Crown.

This stone is one of the earliest that figure in the list of talismans, or gems whose native virtues were heightened by the sigil engraved upon them, a superstition still in its infancy in the age of Pliny, when, although the medical virtues of many gems were generally admitted, the doctrine of their supernatural powers was as yet ridiculed by the learned as a figment of the credulous East. Thus under this head Pliny remarks that "the lying Magi hold out that these gems are an antidote to drunkenness, and take

their name from this property.* Moreover, that if the
names of the Moon or Sun be engraved upon them, and
they be thus hung about the neck from the hair of a
baboon, or the feathers of a swallow, they are a charm
against witchcraft. They are also serviceable to persons
having petitions to make to princes: they keep off hail-
storms and flights of locusts with the assistance of a spell
which these doctors teach." Where it may be remarked
that the names of the Moon and Sun must have been the
mystic names of these luminaries emblazoned in a sacred
Oriental language, for the same in Greek or Latin would
have been too simple to captivate the faith of their dupes,
and besides, never are to be seen on any similar relics.
Doubtless we have here an allusion to the ΙΑΩ and ΑΒΡΑΣΑΞ,
words of that actual signification, and which in the following
century appear on such countless multitudes of talismanic
gems connected with Magian or with Egyptian superstitions.

* A notion curiously illustrated by a four-sided Amethyst pendant
(Blacas) presenting on each face a Bacchante in a different attitude of
frenzy: a fine Greek work.

ASTERIA : *the Girasol Sapphire.*

PLINY thus describes the stone :—"The Asteria, holding
the first rank by a natural claim, because it contains a
kind of lustre within its substance like the pupil of the
eye, and pours it from side to side, when held at an angle,
as though it moved about in the inside, presenting itself
now from one point, now from another. Also, when held
against the sun, it reflects rays of a brilliant white, whence
it derives its name (Star-stone). The Indian sort is hard
to engrave upon. That found in Carmania is preferred to
the others." No description can better suit our Girasol
Sapphire, which in its native state is usually somewhat
globose, being composed of concentric layers like an onion,
to which arrangement its opalescence is due. In this state
the light appears as a small orb shifting to and fro within
the stone, according as it is turned ; but when cut to a
plane and polished, this orb becomes a most delicate star of
five, or six, silky rays diverging from one centre. This
variety is pale and milky : hence the propriety of the term
"candicantes," applied by Pliny to its radiance. The
character of extreme hardness restricts his description to
this kind of Sapphire ; for the Adularian Felspar, or
Moon-stone (see SOLIS GEMMA), answering to it in some
respects, is extremely soft. This latter would indeed
seem to be the next stone on his list, the Astrion, also
colourless, approaching to Rock-crystal in appearance, produced in India on the coast of Patallene : " within from

the centre shines a star with the light of the full moon."
Indeed this exactly applies to the Moon-stone, the Indian
sort of which is incomparably superior to the European,
and which presents upon its silver-grey surface that softly
lustrous and full orb comparable to nothing except the
luminary whence comes its name.

Or again the Cymophane (opalescent Chrysoberyl) may
lay claim to the honour of Pliny's notice : for of all gems
displaying the lunar orb floating upon their pallid ground,
this is incomparably the most beautiful: and coming also
from Ceylon may have had its charms appreciated by the
Roman ladies. It possesses, too, the extreme *hardness* he
specifies.

Plutarch (' On Rivers ') says that the Sangaris produces
the gem called *Aster*, which is luminous in the dark, and
therefore called *Ballen*, "the King," by the Phrygians.
Ptolemy Hephæstion the astrologer describes a gem, Aste-
rites, found in the belly of a huge fish named Pan, from its
resemblance to that god. This, if exposed to the sun, shot
forth flames, and was a powerful philtre. Helen used it
for her own signet, engraved with a figure of the Pan-fish ;
and owed to it all her conquests. Suidas quotes the same
legend on the authority of Æsopus, *reader* to King Mithridates. The superstitious Asiatic evidently did not acquiesce
in Tibullus' axiom—

" Forma nihil magicis utitur auxiliis."

Beauty, resistless, craves not magic's aid.

Some, adds Pliny, derive the name of Astrion from the
stone's property of imbibing and reflecting the light of the
stars, if held opposite to them. The best sort was found
in Carmania, being subject to no defects. The *Ceraunia*
was the same thing but of inferior quality. The worst
kind of all resembled the light from a lantern ; which means

a very dull yellow, for the lanterns of the ancients were glazed with *horn.*

The Astroites, celebrated for its magic virtues by Zoroaster, appears to be the same as the Astrion.

The Astrobolos also may be regarded as a Moon-stone, for Sudines describes it as being like a fish's eye, and emitting a white radiating lustre in the sunlight. The same properties have gained for a variety of the Adularia, the title of ' Œil de poisson,' with modern jewellers.

The *Selenites* has a better claim, from its specified *greenish* tint, than any of the preceding to be considered the Cat's-eye. (*See* ZMILAMPIS.)

The only conclusion to be derived from the brief notices of the several stones above quoted is that *Asteria* and similar names were used in different senses by authors of different times, but that *Pliny* understood by Asteria the same gem as we do at present—the Star-Sapphire; and that this, when exhibiting more purple than blue, was distinguished as the Ceraunia, or Lightning-stone.

The Star-Sapphire appears also to be intended under the title of *Astrapia* (Lightning-stone), " where, in a colourless, or an azure ground, as it were the rays of lightning diverge from the centre.''

BASANITES: *Basalt.*

AN igneous rock, extremely fine-grained, of a deep black, but showing a slight tinge of green when viewed at a certain angle. The Romans got it from Egypt, and its name (Coptic) implied that it had both the appearance and the hardness of iron (**xxxvi. 11**). Indeed, sculptures in Basalt have greatly the appearance of cast-iron figures. The surface, however highly polished, exhibits a granular texture, serving to distinguish it from the rarer antique black marble, the " Nero antico."

The largest work in Basalt known to the Romans was the recumbent Nile, with his sixteen infants sporting about him—alluding to the number of cubits attained by his annual rise. This had been dedicated in the Temple of Peace by Vespasian, but is not the one now in the Capitoline Museum which passes for it, the latter being in *marble*. It was apparently a work of the age of the Ptolemies, and brought from Alexandria by the Emperor named as the dedicator. Pliny mentions a report that there existed at Thebes a similar statue, but of Serapis. The Capitoline Museum in fact possesses some wonderful Centaurs and Stags in Basalt, ascribed to the reign of Hadrian, of whom Pausanias had remarked as something noteworthy a statue in this material, " Egyptian stone," in the Olympeum, Athens. His love for Egyptian art had revived the use of this stone, in which some extraordinary monuments of the earliest times were executed, such as the sarcophagus in the British Museum commonly known as the Tomb of Alexander.

The Egyptians continued to sculpture Basalt, then known as " Marmor Thebaicum," to a late period of the Roman Empire. Spartian describes a statue of Pescennius *Niger* in this stone, a choice allusive to his name, a present from the Prince of Thebes, as then standing in Niger's former palace at Rome, and thus barbarously Latinizes the Greek dedicatory lines below it :—

> " Terror Ægyptiaci Niger astat militis ingens
> Thebaidos socius aurea sæcla volens.
> Hunc reges, hunc gentes amant, hunc aurea Roma ;
> Hic Antoninis carus et imperio.
> *Nigrum* nomen habet, nigrum formavimus ipsi,
> Ut consentiret forma metalle tibi."

The more compact pieces of this extremely hard material were used for scarabei and intagli by the later Egyptians. It is not unusual to find Gnostic amulets, belonging to the Alexandrian sects, engraved in Basalt. Engravers, however, of a good period, have never made use of so coarse a material. The only black stones ever presenting intagli of artistic value are the Black Jasper and Agate, and even they were but in small request with the Romans.

It is unaccountable why the Romans should have considered the Basalt as an exclusively Ethiopian production, when precisely the same rock was being quarried a few miles off in the Campagna (near Baccano), in vast quantities, to pave their roads; the "·silex " of the Via Appia being a true Basalt. Perhaps, however, these suburban quarries did not supply masses of sufficient dimensions to be worked into statues ; and it is still more likely that all the sculptures in Basalt known to the Romans had been executed in Egypt, which would to a certain extent have prevented their discovering what really was the material.

Basalt more probably took its name from its resemblance in colour to the *black* pebble used by the Greeks for testing coin upon, which appears to have been the same substance, and similarly used, as our Touchstone. The French "basané," "swarthy," seems to have the same origin. This βασανὸς is largely dwelt upon by Theophrastus (45-49). The .stones of the kind were only found in the river Tmolus, were round and flat like a counter (ψῆφος), but twice as large as the largest counter; and their upper side, that had lain exposed to the sun, was better for the *touch* than the lower, whence a moisture was apt to exude. By means of this stone the exact amount of alloy, either copper or silver, in the standard of the gold stater could be ascertained. (*See* LAPIS LYDIUS.)

BATRACHITES: *Toad-stone.*

THE brief notice of this stone by Pliny contains no more than, " Coptos also produces the Batrachites, one sort like a frog in colour, another ebony, the third of red mixed with black." Small figures of frogs (done in the later Egyptian style,) perforated to be worn as amulets, occasionally are to be met with cut in full relief out of a yellow and green Jasper, reproducing with singular exactness the actual appearance of the reptile, and which, therefore, has some claim to be pronounced the ancient Batrachites. No further notice of this stone can be traced in the other writers of antiquity.

But this singular epithet, primarily intended only to denote the peculiar colour of the stone, furnished later times with the foundation for a most marvellous fable, which long obtained, as the number of examples still preserved attest, universal credit throughout Europe. Understanding the ancient term as implying the natural production of the animal according to the analogy of other similar names, as the Saurites, Echites, &c., doctors taught that the " toad, ugly and venomous, wears yet a precious jewel in his head." A full account of this will be found in that repertory of mediæval medical lore, the ' Speculum Lapidum ' of Camillo, who, as physician to *Borgia*, ought to know something about poisons. He describes it by the names of Borax, Nosa, and Crapondinus, and as being found in the brain of a newly-killed toad. There are two kinds, the

white, which is the best, and the dark with a bluish tinge
with the figure of an eye upon it.* If swallowed it was a
certain antidote against poison, in its passage through the
bowels driving out all noxious matters before it. More
than a century later Vossius asserts (De Phys. Christ. vi.
19) that it was usual to take the Bufonites (Toad-stone) in
drink before meals, to counteract any poison that might be
administered in the dishes; a singular dinner-pill, exemplifying the very uncomfortable state of society in those times.
It was also believed to burn the skin, at the mere presence
of poison, if worn set open in a ring so that the stone
should touch the finger; besides which, it was good against
all complaints of the stomach and kidneys if so carried.
For these virtues, says De Boot, it is much worn in rings, in
spite of its ugly colour: a fact which innumerable examples
remain to confirm. Nevertheless, this invaluable guardian
sold cheaply enough: "the price asked by the vendor being
regulated by the eagerness of the purchaser to possess it."
Chinese porcelain also in the times of Vossius was supposed
to fly into pieces when a poisoned draught was poured into
it. Erasmus, in his 'Peregrinatio Religionis ergo,' thus
describes a famous Toad-stone dedicated to Our Ladye
of Walsingham:—"At the feet of the Virgin is placed a
gem to which no name has yet been given among the
Greeks or Romans, but the French have entitled it after the
toad, inasmuch as it represents the figure of a toad so
exactly that no art of man could do it as well. And the
wonder is so much the greater, that the stone is very small :
the figure of the toad does not project from the surface, but
shines through as if enclosed in the gem itself. And some,
no mean authorities, add that, if the stone be put into
vinegar, the toad will swim therein and move its legs."

* The brown marks, seemingly produced by oxide of iron, observable
on some of these fossils.

This account makes it probable that the gem in question was a lump of amber enclosing some large insect.

Another version was that in order to preserve its virtues this jewel must be voluntarily surrendered by the living reptile. De Boot relates how that in his boyhood he had sat up a whole night watching a toad placed upon a red cloth (the received mode of making it disgorge the treasure), in the hopes of seeing it cast up the Bufonites; but the experiment having failed, he concludes the story to have no foundation in fact. The figures he gives of the substance leave no doubt that he had in view the small fossil hemispheres, like nutshells, found in the Green-sand formation in considerable abundance. It is probable enough that the toad, like certain larger reptiles (the alligator),* may be in the habit of swallowing small stones to assist digestion, and such, if found by accident within an animal enjoying so great a reputation in the Middle Ages for his powers both for good and evil in medicine and witchcraft, would naturally be taken for a most potent gem.

Many Toad-stones are still preserved in collections of mediæval jewels, set in silver rings, the metal appropriate seemingly to that purpose, being generally prescribed for the setting of all amulets. I have lately had an opportunity of examining several of these Toad-stones, some in their original settings, some extracted. They are hemispherical, elliptical, or oval, hollow within, of an apparently petrified bony substance, whity brown, or variegated with darker shades.† The true, but very recent explana-

---

* In fact the 'Cayman-stone' found in the belly of the alligators caught near the town of Nombre de Dios are mentioned in his list of medicinal stones as greatly valued both by the Indians and Spaniards as cures for the ague, one stone being applied to each temple of the patient.

† These darker shades often produce a 'nature-picture,' which to a fanciful eye already prepared by the name of the stone, presents a lively image of the mysterious reptile and thus endorses its received character.

tion of their origin is, that they were the bony embossed
plates lining the palate or the jaws, and serving instead of
teeth to a fossil fish, an arrangement observable in the
recent representatives of the same species.*

Pliny indeed includes in his alphabetical list (though
with frequent sneers at the credulity or roguery † of his
authorities) numerous " stones of virtue " generated within
the viscera of various animals, viz., the Alectoria, like a
crystal, as big as a bean, found in the belly of a cock, " by
the virtue of which they pretend that Milo of Crotona
gained his reputation for being invincible." Mediæval
naturalists added that this stone prevented the wearer
from ever feeling thirst, and that the *capon* containing the
fatal treasure was betrayed by his never drinking with his
food like other fowls. There is no doubt that crystalline
stones are often found in a cock's gizzard, as fowls pick up
anything shining in preference to other gravel, for which
reason in Brazil their gizzards sometimes contain diamonds.
The sole diamond ever discovered in Europe, namely, that
obtained some years back in Wicklow, Ireland, is supposed
by Prof. Miller to have been thus transported thither by
some bird of passage. It was picked up in the bed of a
rivulet, and sold, instead of a pearl, to a resident lady, who
was accustomed to buy the mussel-pearls from the finders
at the fixed rate of one penny each. This theory of its
transportation bears upon the singular fact that the mi-
gratory pigeons caught at certain seasons about Gibraltar
are sometimes found to have particles of gold-dust
sticking to their feet, reminiscences of their scrapings

---

* I have at last had the satisfaction of verifying this by the inspection of the part of such a jaw-bone from the Green-sand formation,
still thickly studded with many such bosses, some the identical hemispheres in size and nature to be seen set in rings, diminishing down
into others no larger than a mustard-seed.

† " Magorum mendacia."

amidst the unapproachable treasures of the centre of the Sahara.

The Brontia (supposed to find its way into the head of the tortoise during thunder-storms), extinguished fires occasioned by lightning.

The Cinædia, found in the brains of the fish so called, of a white colour and oblong shape, foreshowed tempests by a change of hue, being clear in fine, but cloudy in stormy weather.

The Chelidonia was of two kinds, in colour like the swallow, on the one side entirely purple, on the other purple dappled with black spots. But Isidorus, improving upon this simple statement, which derives the name merely from the colour, makes these stones to be found *inside* the swallow, of two kinds, white and red, of wonderful virtue for the cure of insanity and for giving good fortune to the wearer.

The Chelonia, or Indian tortoise's eye, smeared with honey and laid on the tongue, bestowed the power of divination for periods varying according to the moon's age.

The Chelonitis resembled in shape the tortoise; the Magi told wonderful stories of its power in appeasing storms; nevertheless the kind starred over with gold spots, if thrown, together with a beetle, into boiling water, would raise a tempest.

The Chloritis, of a grass-green, was to be found inside the wagtail; this the Magi recommended should be mounted in an iron ring "to produce some of those prodigious effects they are wont to promise." The Draconitis or Dracontia, from the serpent's brain, had no lustre unless cut out of the living reptile. Sotacus, who had himself seen it, says it was colourless and transparent, could not be polished, or admit of any artificial improvement. In

this quotation from Sotacus (elsewhere styled by Pliny
" e vetustissimis auctoribus ") we find the first allusion to
the real Diamond and its two distinctive characters, as
well as to the popular Indian tale of the Valley of Serpents
(known to us all from Sinbad's adventure), that alone
produced it. Sotacus adds that the hunters went in quest
of it in chariots, and when pursued by the serpent fled,
strewing the ground in their flight with some soporific
drugs by which the reptile was overpowered, whereupon
they turned again and made prize of his head.

Philostratus, however, gives fuller details of the mode
of capture (iii. 8). " These dragons are thus taken : the
Indians having woven letters of gold into a scarlet robe,
spread it out before the den, but first of all magically
infuse a soporific power into these letters, whereby the
dragon hath his eyes overcome, losing all power to turn
them away. They also sing over him many spells of their
mystic art, whereby he is drawn forth, and, putting his
neck outside of his den, falls asleep upon the letters.
Then the Indians, assailing him as he lies, cut off his head
with their axes and make prize of the gems within it,
for in the heads of these mountain-dragons are secreted
gems bright-coloured to the eye, and reflecting all kinds of
hues, of virtue moreover indescribable, like the ring of
Gyges. Often too doth the dragon seize the Indian, axe,
charm, and all, and escape with him into his hole, all but
making the mountain tremble."

The Hyænia, existing within the eye of that beast,
which was hunted for its sake, being placed under the
tongue conferred the gift of prophecy ; and lastly, the
Saurites was to be procured out of the belly of a lizard cut
open with a knife made of a sharp reed.

Pliny concludes his alphabetical list with this not un-called-for observation: "There are many stones besides these, and of an even more prodigious description, to which writers have given foreign names, allowing them to be minerals (lapides), and not gems; it will be enough for us to have, in the foregoing list, exposed a few of their awful lies" (dira mendacia).

BERYLLUS: Βήρυλλος: *The Beryl.*

PLINY notices (xxxvii. 22) that many were of opinion that
• this stone was of the same or a similar nature with the
Emerald. This opinion has been proved correct by modern
analysis, the component parts of each, in the same proportions, being Silica, Alumina, and Glucina, coloured by the
oxide of Chrome. On these grounds, Phillips (Mineralogy :
Beryl) states that "the only important difference between
Emerald and Beryl is their colours; which, since they
present an uninterrupted series, is altogether insufficient
for a division of the present species." The Emerald is
distinguished by its peculiar "emerald-green," which it
derives from a small proportion of Chrome: all the varieties
of other colours, tinged more or less yellow, or blue, or
altogether colourless, are Beryls. Yet, though thus identical in their chemical constitution, the Beryl is by far the
harder of the two, its hardness being denoted in the scale
by 7·5 to 8, and therefore even superior to that of the
Garnet. It appears also to be of a singularly compact
texture, for antique intagli in Beryl are found retaining
their original surface-polish more perfectly than those in
almost any other material. Not so the Emerald.

This gem held the same degree of rarity amongst the
Romans as the true Smaragdus itself, for it was then
obtained from India almost exclusively, "being rarely
found elsewhere" (Plin). Dionysius Periegetes enumerates it amongst the gems gleaned by the Indian Ariani

from amongst the pebbles of their torrents, and in another passage as met with embedded in serpentine rock in Babylonia. Pliny distinguishes its varieties with all the exactness of modern science. "The most admired," says he, "emulate the green tint of pure seawater" (the present Aquamarine). "Next in favour is the Chrysoberyl, in which this green is tinged with a golden lustre" (an exact description this of our Indian Chrysolite, as the jewellers invariably term it, although mineralogists retain for it its ancient name of Chrysoberyl. In its composition, alumina constitutes 70 to 75 per cent., the glucina being greatly diminished in proportion; the stone is therefore much harder than the true Beryl). There was a still paler kind known then by the name of Chrysoprase, but in fact only a variety of the last mentioned. Then came the Hyacinthizontes, or sapphire-like sort; and the Aëroides, of a yet fainter shade of sky-blue.* The last was the species invariably employed for intagli by the ancient engravers, as all existing antique examples attest. Lowest of all were ranked the wax-coloured, the oily, *i. e.* those of a greasy yellowish green, and such as were totally colourless.

The Beryl was the only one amongst the precious stones that was facetted by the Roman jewellers, who cut it into a sexangular pyramid, as otherwise it had no brilliancy. Beryls were then highly prized both for the purpose of ear-drops and of mere ornamental, *i. e.* not engraved, ringstones. When Cynthia's shade appears to Propertius he remarks that (iv. 7)—

" Et solitam digito beryllon adederat ignis."

---

* St. Laurent thinks this was the true Sapphire, and Lessing agrees with him (Ant. Briefe xxvi.), asserting that the ancients knew the Sapphire only as a sort of Amethyst, or of Beryl. But this is an

And Juvenal's ostentatious host Virro drinks from a bowl embossed with Amber reliefs and Beryl—

> " Heliadum crustas et inæquales Beryllo
>   Virro tenet phialas."

Inasmuch as its chief value consisted in the length of the native crystals, the Romans, according to Pliny, preferred to make " Cylindri" or pendants out of them rather than ring-stones. Such were the costly presents that, as Juvenal hints (ii. 61), would reward the bride for her discreet silence—

> " Tu nube atque tace, donant arcana cylindros."

The Indians cut this stone similarly into long cylindrical beads, and wore them strung upon elephant's hair, believing that their lustre was heightened by this perforation and the removal of the internal substance, or " paler marrow." But the most perfect in colour were not bored, but used for decoration by having each end secured within a gold boss.

This nation had also the art of tinging the common Rock crystal so as to make it pass for the Beryl : probably by plunging it when heated into some menstruum saturated with oxide of copper ; as the *Rubace* of the French is still produced by thus treating a piece of Crystal in a spirituous solution of cochineal. Even now the Indians paint the back of every coloured gem they set, so as to improve the fainter tinted ; for which reason they never mount them in their jewelry without a backing. From this deceptive practice of adding a fictitious beauty to their gems, those in native Indian ornaments are rarely, when taken out, found to be of much value : all of high intrinsic worth are

error, for the Blue Beryl, though much resembling the former, differs totally from it in hardness, lustre, and other characters, which the ancients knew well enough how to appreciate.

sold for the European market, the inferior samples when thus painted being deemed good enough for the native jewellers' work.

Antique engravings in Beryl are almost as rare as in the true Emerald ; * but with this difference, that intagli in the former stone, as far as my experience goes, belong to an earlier period of the art, being for the most part fine work of the "perfect" Greek school, whereas intagli in the Emerald are invariably in the later Roman style, dating from Hadrian's time and his immediate successors. To quote a few of the finest, where too the material is hardly to be distinguished from the pale Sapphire. The earliest is the Taras (or Icadius) on the dolphin, in the (former) Praum Collection, the design of which is placed by Winckelmann in the first class of Etruscan work; a head of Proserpine in the purest Sicilian Greek style, and a hippocampus (Marlborough Gems); and amongst the best specimens of Roman date, the young Hercules, inscribed ΓΝΑΙΟΣ, the patron-god, and favourite *prænomen* of the Scipios (Blacas), with a most elegant design of Cupid borne over the waves by a dolphin (Cracherode). The grandest intaglio extant of the Roman period is also upon an Aquamarine of the extraordinary magnitude of $2\frac{1}{2} \times 2\frac{1}{4}$ inches; the bust of Julia Titi, signed by the artist ΕΥΟΔΟC ΕΠΟΙΕΙ. For nearly a thousand years it formed the knosp of a golden reliquary celebrated as " l'escrain de Charlemagne," presented by Charlemagne to the Abbey of St. Denys, in which it was set with the convex back uppermost, being regarded as an invaluable Emerald. The rarity of such Beryl intagli, when of unquestionable antiquity, is readily

---

* This stone has been the vehicle for carrying down to us the fame of one of the only five gem-engravers noticed by ancient writers now extant. Addæcis (the court poet of Polemo, who was made King of Pontus by Antony), has immortalized a work of " Tryphou's " in Beryl, the Nymph Galene, in a very graceful epigram. (Anth. ix. 544.)

accounted for by the extreme value this, then so rare
material, bore amongst the ancients, equal in fact to that
of the true Emerald : but its modern plentifulness, coupled
with its beauty, has rendered it a favourite stone with the
artists of the Renaissance and of succeeding times. These
recent works may be distinguished as invariably executed
upon the green sort or Aquamarine ; the Aëroides or sky-
blue of the ancients either not being any longer obtainable, or else of much greater rarity. It is the vast supply
poured in from Saxony, Siberia, and America, that has
sunk the value of this beautiful stone so low in modern
times. It possesses very great lustre, especially by lamplight ; for which reason the lighter-coloured varieties have
long been used in jewelry as fraudulent substitutes for the
true Diamond, a deception noted as commonly practised in
the fifteenth century by Camillo when treating of the
tricks of the jewellers in his own day.   At present it
is similarly employed in Germany under the name of
" Diamond of the Rhine." In consequence of this substitution, it happens that people have often flattered themselves with being the owners of a Diamond of enormous
value, which, on examination by a skilful lapidary, has
turned out to be nothing more than a small, worthless
Aquamarine. The stone has gone completely out of
fashion in this country (though not in Italy) ; the natural
result of the profusion in which it is now produced by
the different regions above mentioned, and that too in
masses often of enormous size, their dimensions reminding
us of the monstrous smaragdi spoken of by Apion and
Theophrastus. Remarkable specimens in the British
Museum are two Beryls from Ackworth, New Hampshire,
one weighing 48, and the other 83 pounds.
     The most singular modern work ever executed in this
stone is the sword-hilt of that brilliant fop, Murat, of

one monster facetted aquamarine, with a big jacinth for
its pommel, and the ends of the cross-guard terminating in
lions' heads carved in diamond.

It is a curious fact that *Beryllus* is the Low Latin term
for a magnifying glass: hence the German "Brille," a pair
of spectacles. For this reason Nicolaus de Cusa, bishop of
Brixen (who died 1454), gave the name of "Beryllus" to
one of his works, "because by its aid the mind would be
able to penetrate into matters which otherwise it would
be unable to pierce."* Which designation he thus explains
in his second chapter, saying, "the Beryl is a shining,
colourless, transparent stone, to which a concave as well as
a convex form is given by art; and looking through it
one sees what was before invisible." Probably the first
idea of this invention was caught by accidentally looking
through a double-convex and clear Beryl, or one cut "en
cabochon" (the usual form given to antique transparent
gems), and thence concluding that a piece of glass similarly
shaped would produce the same effect in magnifying minute
objects. Mediæval glass being never colourless, but always
tinged more or less with green, the resemblance as to colour
and form of a lens in such a material to an actual Beryl
was sufficiently obvious to induce the communication of the
name to the new discovery. Protoprodromus, as early as
1150, humorously describes the physicians of Manuel Comnenus as using the ὑέλιον to examine the nature of his

---

* A similar derivation exists in the French language, where "loupe,"
a magnifying or convex lens, comes from the Low Latin "lupa," an
unpolished precious stone. The term seems to have been at first a
simile allusive to the untamed condition of the stone. It appears
restricted to the Sapphire, a gem more or less convex in its native
form. Thus, in the Jocalia Sti. Thomæ (Dart. Hist. Cant. Cath. Ap.
xiii.)—Annulus magnus cum Sapphiro nigro qui vocatur lup.—It. Annulus cum parvo Sapphiro nigro qui vocatur lup." That "lup" denotes
the shape, appears from the next item, "annulus cum Sapohiro quadrato
aquoso."

evacuations: and this Lessing supposes very plausibly to mean a magnifying lens (Ant. Briefe, xlv.)—

" Θεωροῦσι καὶ τὰ σκόβαλα μετὰ τοῦ ὑελίου."

Now this is the very term ὕαλος used by Socrates to describe a burning glass (and consequently a magnifying lens) in Aristophanes (Nubes, 758). By some lucky accident the observer of this property in his Beryl had been led by induction to apply a fact, similar to that involved in Nero's recorded use of his Emerald lorgnette so many centuries before, to the working out of a most important result, through the happy thought that the marvellous effect was due not to the occult virtue of the gem itself,* but to the artificial shape imparted to it.

In the absurd nomenclature current with the English lapidaries in the last century, as Lessing has noticed, the name of Beryl was given to every variety of the Sard in which yellow predominated : the red alone, following the French example, was distinguished as the Cornelian. Dr. Woodward and Hill both notice this singular misnomer. In the same jargon the true Beryl is only mentioned as the Aquamarine. Even Natter, who should have known better, has adopted the same system of misnaming these stones, in the various Catalogues he has drawn up. This has been a fruitful source of error to foreign archæologists, who, trusting to the English description, give so large a proportion of works preserved in our Cabinets as upon Beryl which are in truth on Sard, according to the well-known rule of the preponderance in numbers of the latter over all other species, used by the ancients, collectively.

---

* The concave Emerald was supposed to aid the myopic eye, because its nature was beneficial to the sight.

CALLAIS *and* CALLAINA : Καλλαϊνὸς λίθος :
Turquois: Peridot?

OWING to a corrupt reading in the old editions of Pliny
(xxxvii. 33) of " Callais " for " Callaina," the former name
is now universally supposed to have anciently designated
our Turquois. This identification, however, seems to me
far from being borne out by Pliny's description of the most
striking characters of the stone in question. "Next to the
Topazios (Peridot) in appearance, though not in value,
comes the Callaina, of a pale yellow mixed with green.*
Its native country is the region to the north (post aversa)
of India, amongst the tribes of Mount Caucasus, the Phy-
cari, Dahae, and Sacae (Little Bokhara). It is of extraordinary magnitude for a precious stone, but full of flaws
(fistulosa) and of dross. A clearer and better kind is the
one obtained from Carmania. In both countries it is found
adhering to the surfaces of the rocks, and protuberant
therefrom, after the manner of an eye. These rocks being
inaccessible, the natives knock down the gem, with all the
moss that surrounds it, by means of bullets thrown from

---

* Pliny's peculiar term for its colour, "e viridi pallens," can only
denote a preponderance of pale yellow over the green. For *pallidus*
(like χλωρὸς, in speaking of the complexion) signifies a pale yellow,
as clearly is shown by the comparison, "pallidior buxo." Pliny's
Topazios, on the contrary, is "in suo genere virens,' for here the green
was the predominant tint. Another proof, added to those remarked
below, that he is treating of mere varieties of the same material.

slings at a distance. This gem constitutes their wealth,
and their choicest ornament for the neck and the finger;
their chief glory, the number each man has knocked down
in his lifetime, success in the pursuit being quite a lottery,
sometimes the very first shot has brought down several
beautiful stones; others again have tried for them till old
age, without obtaining a single one. The stone is shaped
by a cutting instrument (sectura), being too brittle to be
worked in any other manner. The best have the colour of
the Emerald, whence it is plain that they please with a
beauty not their own. No gem is more improved by setting
in gold, and gold itself is by no gem so well set off. The
finer sort lose their colour through wetting with oil, grease,
or wine; but the inferior retain theirs more permanently.
No other stone is so easily imitated in deceptive paste." *

All these particulars indicate a pale-green *transparent*
stone, in short, an inferior Peridot, or rather the Peridot
itself, as distinguished from our jewellers' Chrysolite, to
which it yields considerably in point of hardness. The
expression "full of holes and of dross," is quite inconsistent
with the idea of an *opaque* solid body like the modern
Turquois, the generic distinction of which would, besides,
certainly not be described as a " pale-green " colour. The
Peridot also exhibits the same " extraordinary magnitude,"
as compared to other gems; whilst, similarly, from its softness, it is extremely difficult to polish. Its colour, too,
evaporates through long exposure to the light. De Boot
conjectured that our Aquamarine was intended by the
Callaina, but the superior hardness of the former is of itself
sufficient to overthrow such an explanation. The notice of

* " Mendaci vitro." Nothing resembles the Peridot so closely as a
thick fragment of green bottle-glass. Orpheus (277) actually uses the
epithet ὑαλοειδέες τόπαζοι, "the glass-like topazi." Light-green transparent pastes are the commonest among the antique; imitations of the
blue, opaque Turquois, on the contrary, do not occur.

the Callaina to be found in Solinus is to the same effect as that cited from Pliny, but with the important addition "that it comes next to the Emerald in price and estimation :" a further support of my own idea of its real character.

Salmasius was the first to conjecture that "Callaina," not "Callais," was the true reading in this chapter of Pliny's,—a correction since verified by Jan's discovery oɪ the genuine text,—and supposed it connected with the name of a peculiar green dye, "Callaicum." Amongst the Indian exports specified in the 'Periplus of the Red Sea' we find "pepper, the stone *Callainos*, Lapis-lazuli, indigo." This being the case, the Callaina may reasonably be supposed a variety of the Topazios, but derived from India and Persia, whereas the original and finest kind was considered peculiar to the island of the Red Sea, where it had been first discovered (TOPAZIOS). For Peridots vary very greatly in colour. Those of the deepest green are naturally the most esteemed : others again are all but colourless. Such a distinction was quite sufficient to oblige the Greek mineralogists, judging solely by the eye, to class the Topazios, deep-green, and the Callaina, yellowish-green, under different species.

The *Callais*, however, which is entered separately in Pliny's alphabetical list of the inferior stones (c. 56), and therefore must have been regarded by him as something totally distinct from his *Callaina* (a gem of sufficient importance to engross an entire chapter to itself), has much better claims to stand for our Turquois ; for " it resembled Lapis-lazuli, only was whiter (candidior), and of the tint of the sea where it is shallow."

The older mineralogists, like De Boot, took Pliny's "Jaspis aërizusa," or cerulean Jasper, for the Turquois ; but without good reason, that stone being incontestably our

Sapphirine Calcedony (JASPIS). On the other hand, there are many points of analogy to induce a belief that our Turquois was the Hermean or Persian Smaragdus of Democritus, "protuberant, *not transparent*, of an agreeable and even colour, like the eye of a leopard or cat; losing their beauty in the sunshine, but lustrous in the shade, and conspicuous from a greater distance than any other precious stone :" one grand distinction of the Turquois being that its colour is improved by artificial light, in which, even though greenish, it becomes a perfect celestial blue. A strong confirmation of this view is to be found in the war so hotly waged between the mediæval doctors as to whether that great oracle of the science, the Arabian Mesues, meant the Emerald or the Turquois by his "Firuzegi,"* that essential ingredient in his highly-esteemed electuary.

The "Fossil Ivory" of Theophrastus (37) " dark blue mottled with white," can hardly be other than the Occidental Turquois, or petrified bone, tinged by copper oxide, in which the osseous structure is plainly discernible under the microscope, and which, besides, is much softer than the true Persian Turquois ("de la vieille roche" in jewellers' parlance), which is sufficiently hard to strike fire with steel. According to Hill, the blue that mottles the surface of the former species can, by the judicious application of heat, be evenly diffused throughout its mass, thus greatly improving its appearance, and often enabling it to pass muster for the actual and precious *stone*. Its Greek appellation, so much more correct than the modern, may be set down amongst the proofs that continually strike us of the sagacity of the primitive naturalists in discovering by observation and induction alone the true origin of many substances. Barbot found by experiment that the teeth of

* "The Victor," an epithet attesting its reputation amongst the Persians.

certain graminivora, by immersion in a strong alkaline solution of sulphate of copper, are speedily converted into a regular Occidental Turquois, more especially as regards their enamel.

That our Turquois, *i.e.* "Turkish-stone," a name containing in itself an analogy to the antique "Persicus" Smaragdus, is in truth no other than that stone, is to a certain extent deducible from Theophrastus' notice of the special employment of the latter in the λιθοκόλλητα, or jewelled plate of his own times. Although no specimens of that remote antiquity have come down to us, we may safely conclude from the invariableness of Oriental usages that the Persian inlaid jewelry (the only make known to the Eresian philosopher) was then, as now, set, or rather incrusted, with no other gem than the Turquois. When enamels came into fashion under the Lower Empire, their predominating blue was doubtless intended to represent economically the more precious natural production. Indeed, the Merovingian gold dish, known as the "Lanx of Gourdon" (Bib. Impériale), probably a Byzantine work, displays in each corner an inlaid heart of Turquois.

The Turquois became known to the Greeks by its occurrence in the spoils brought home by the Macedonian soldiers from their Persian campaigns, the goblets, dishes, and armour of gold—over all of which this lovely gem was as profusely lavished by the ancient Achæmenidæ as by their modern representatives, whose splendour in this particular struck the jeweller Chardin with amazement when he was permitted to inspect the Shah's treasure-house. To give one example only from antiquity : the sheath by itself of the sword of Mithridates (when his corpse in its royal attire was sent to Sylla) was valued at 400 talents (80,000*l.*). One Publius purloined this piece, and sold it afterwards to King Ariarathes.

This identical art of inlaying with Turquois makes its appearance again in a quarter where least to be expected— amongst the Aztecs, when first discovered : wood or bone, however, being substituted for gold in the groundwork. Longolius gives a minute description of a monument of Mexican ingenuity in this line, shown to him by Gri- moaldus, Charles V.'s envoy to the Pope. He took it for a mask of the Devil (classically expressed. as "persona Furiæ"), and, in truth, it fully bore out that idea. It was of wood, the features being represented by an inlay of Tur- quois, Emerald, and green and red Jaspers, as artificially as if painted; the eyes of red amber, reflecting the beholder's face in a terrific manner. In the Hertz Collection was a mask of the same kind, but less elaborate, the inlay con- sisting of Turquois alone, with mother-of-pearl eyes, and tusks of ivory, the whole being "of aspect very hideous and terror-inspiring." It was accompanied with the sacri- ficial knife of Obsidian. Such an appropriate disguise is said to have been assumed by the priest of Tlaltoe, god of the earth and lower regions, when about to offer to him the accustomed human sacrifice on the summit of the Teocalli. Another relic, in the same collection, of this art was a *human skull,* similarly inlaid with patterns in Turquois and Obsidian ; and the sockets of the eyes filled in with disks of mother-of-pearl.

But to return to Grecian mineralogy, we need not be surprised at finding the term "Smaragdus," the synonym for *green,* applied to a gem now only valuable when of a pure *blue.* De Boot in his day defined the best as "exactly the colour of purified verdigris, presenting to the eye the agreeable greenness of a milky blue :" and even in the last century, according to Dutens, many people preferred the Turquois of that tinge, if pure and unclouded. The very rare antique works in relief in Turquois, which have come down to us, are all executed in the green sort; the principal being the bust of Tiberius (Florence), the head as large as a walnut, sculptured in full relief; and the busts of Livia and the same emperor as a child, in half relief, on a stone of much larger dimensions (Marlborough)—the latter a carving of extraordinary merit. The last-named cabinet also boasts a cameo of still greater rarity—a small portrait of a Greek prince, in a Turquois beautifully azure, and evidently "de la vieille roche" in every sense of the term. Antique *intagli* in this stone, owing probably to its want of hardness, do not exist at all, except a few examples in the Sassanian class. But the Renaissance artists employed it largely for small heads, "en ronde bosse," and yet more for camei; such being the true origin of almost all these small works in Turquois, though usually regarded as antique. The modern Persians practise the converse art to that of their ancestors, the λιθοκόλλησις: instead of inlaying gold with Turquois, they inlay Turquois with gold: engraving cyphers and arabesques upon its surface, they fill them with fine gold, beaten in after the manner of damascening steel. Considering the fragility of the substance, how this is effected remains a mystery to European lapidaries.

This stone has pretty much the same *chemical constituents* as the Lapis-lazuli, but seems to possess a somewhat softer paste; and, for the most part, becomes decomposed and chalky by long lying in the earth, as is often to be remarked in the Turquois set in jewelry so discovered though but two or three centuries old.

Ben Mansur states that the best Turquois came then (as now) from Nishapur in Khorasan, the rest from Ghasnah, Irak, and Kerman. Of the first sort he makes seven varieties, according to their relative degrees of *hardness*, and their tints; that named after Abu-Ishak being the finest, and the Andelibi, of a milky blue, the weakest. His fourth sort, the Sermani, "having spots of gold," can be no other than a shade of the Lapis-lazuli. As to the natural properties of the Turquois, he avers "that it grows clear or dull according to the state of the weather, and on rainy days has a greater circumference than on fine. One species gains a better colour by maceration in oil, but loses it again after a time. According to the age since it has been dug, the Turquois is divided into that of the *Old*,* and of the *New* mine : the latter does not keep its colour." This distinction is still in force; Turquois "de la vieille roche," however, actually means at present nothing more than one of the best quality, and *supposed* not to be liable to fade. Nishapur continues to furnish the only Turquois of commercial value : they are brought stuck upon rolls of wax by Tartar and Persian merchants to the great fair of Nishni-Novogorod. Lately, some have been brought direct to this country from the foot of Mount Sinai, where they occur in strata of a sandstone rock. This new find is of a remarkably rich azure, superior when recent to the finest Persian, but (a fatal drawback) liable to change yet more rapidly and capriciously than any of the old.† The above-quoted

---

* This "Old Mine or Rock," says Chardin, "was in a mountain, Firuz-cos, distant three days' journey from the Caspian. All its produce was reserved exclusively for the Shah; whatever found its way into the market was due to the embezzlement of the miners and their directors; a source of supply by no means scanty. No wonder, therefore, that when Chardin was allowed to visit the Treasury at Ispahan he beheld in each chamber the stones in the rough, piled high on the floor like heaps of grain, and the polished filling innumerable leather bags, weighing 45 to 50 lbs. each."

† The unfortunate liability of so costly a stone to lose its colour and therewith its value, exercised the ingenuity of the lapidaries of De Boot's age in devising methods for restoring it. Some, distilled ultramarine in aqua-regia, and polished the spoilt Turquois with the residuum, others, after macerating the gem in aqua-regia, and washing it in vinegar, threw it into ice-cold water. A third and more approved method was to eat away the surface with oil of vitriol; but of all the recipes (the

Mexican remains confirm Garcilasso de la Vega's statement
that the country possessed the Turquois as well as the
Emerald and the Pearl, and show that its mines must still
exist in some undiscovered locality of the ancient Aztec
empire. It does not, however, appear that the Spaniards ever
came to the knowledge of them; otherwise the high value
of the gem in the age of the Conquest would have rendered
them as much an object for exploration as those of the
Emerald itself.

Few gems were invested with more wonderful properties than the Turquois by the credulity' of mediæval
naturalists. A long list of them is given by De Boot, who
vouches for some of the strangest upon his own experience.
For example, besides strengthening the eyes and cheering
the soul of the wearer, it took upon itself the consequences
of any fall that might happen to him, by cracking itself,
saving him the fracture of a bone. It grew pale as its
owner sickened, lost its colour entirely on his death, but
recovered its full beauty if placed again on the finger of a
new and healthy proprietor. Suspended from the thumb
by a thread and over a drinking-glass, it told the hour by
the exact number of strokes against the sides: "a thing
which, if not occasioned by the pulsation in the holder's
hand, is unquestionably due to the Devil," remarks the
puzzled old physician.* But let him speak for himself;
nothing can be more characteristic of his age :—" The Turquois is believed to strengthen the sight and spirits of the
wearer, but its chief commendation is its protective influence against falls, which, as everybody is assured, it takes

multitude whereof attests their inefficiency) the mechanical one recommended by Tollius appears to me to contain the fairest prospect of
success, viz., to file away the pale surface of the Turquois with emery,
and then to repolish it with tripoli powder.

* Nowadays, a shilling similarly brought into play, will act the
same part of a time-teller; *experto crede.*

(G) F

upon itself, so that the wearer escapes all hurt—a property beyond the scope of reason. I can solemnly affirm that I always wear one set in a ring, the nature of which I can never sufficiently admire. Thirty years ago it had been worn by a Spaniard living not far from my father's house. On his death, when his goods were sold off (as is the custom with us), amongst the rest the Turquois was put up. No one, however, would bid for it, although many had come to buy it on account of its choice colour in the lifetime of the former owner; for it had entirely lost its original beauty and lustre, so that it looked more like Malachite than a Turquois. My father and brother were present, thinking to bid for the gem, which they had often admired aforetime, and were astonished at the change. My father, however, purchased it for a mere trifle, because everybody thought it was not the same that the Spaniard used to wear. When my father came home, thinking it scorn to wear so unsightly a gem, he made me a present of it, saying, 'Since the story is that a Turquois, to exhibit its power, must be presented when one is at home, I now make you a gift of it.' I took the gem to an engraver to cut my arms upon it, as is done upon Jasper, Calcedony, and other cheap stones, not choosing to wear it, having lost all its beauty, merely as an ornament. I received it back from the engraver, and wore it for a signet-ring. Hardly was it on my finger a month· when its original colour returned, though not so bright as before, in consequence of the engraving and the inequality of its surface. Everybody was surprised, more especially as the colour grew finer every day. Perceiving this, I never took it off my finger, just as I do still. Its wonderful virtue in the case of a fall (if really proceeding from it) I have myself experienced. For returning on horseback to Bohemia from Padua, where I had taken my Doctor's degree, a guide I had hired to show me the way points out a footpath by the side of the high road, which I take, and ride on some time in the dark. Suddenly my horse stands still, and will not move a step. I call my guide: he says there is a well in the way, and that I must go back, the path being very narrow. In turning my horse he stumbles, and puts his left foot outside the path, towards the high road. Immediately I feel myself falling, I throw myself out of the saddle upon the road, which was at least ten ells lower down than the path. I fall on my side, and my horse on his back close to me. The guide, not hearing me cry out or speak, thought me crushed to death by the horse; but I was safe and sound, had suffered no harm at all, get on my steed, and pursue my journey. But next morning, as I was washing my hands, I found my Turquois split, and about a quarter of its substance separated from the rest. I therefore got the larger portion of the stone reset, and continued to wear it some years. One day in attempting to lift out of a river, with a long pike, a weight beyond my strength, suddenly the bones of my chest cracked as though a rib were broken, and I felt a dull pain in the side. Thinking something was fractured, I examined, and discovered that the lowest rib was displaced, and its end pushed under the last but one. As the pain was slight, I applied no remedy to the part affected; but the same day, to my surprise, I see my Turquois again broken in two, the smaller portion, however, being no bigger than a hempseed; but lest it should drop out, I had the larger portion, retaining nearly all my arms, set in another ring, which I still wear constantly."

With the Germans it is yet the gem appropriated to the ring, the "gage d'amour," presented by the lover on the acceptance of his suit, the permanence of its colour being believed to depend upon the constancy of his affection.

Inasmuch as this stone is almost as liable to change, and as capriciously as the heart itself, the omen it gives is verified with sufficient frequency to maintain its reputation for infallibility.

De Boot's statement of the fashionableness and value of the stone in his times illustrates Shakespeare, where he makes Shylock say he would not have lost his Turquois ring "for a whole wilderness of monkeys." The lover's custom just noticed must have then prevailed, for he "had it of Rachel when he was a bachelor."

So protective and sympathetic an ornament, a true "decus et tutamen" to its wearer, was with good cause sought after above all other gems. De Boot remarks that no gentleman in his day (ladies, for obvious reasons, eschewing its use) thought his hand becomingly decorated, or his elegance complete, without the adjunct of a handsome Turquois. A stone, the size of a large pea, used to sell for 10 thalers; but if it attained to the magnitude of a hazel-nut (the utmost limit then seen) it was worth 200, or even more.

There are some grounds for the conjecture that the much-disputed Callaina may have been the softer species of the Chrysolite, now termed from its colour the *Olivine*. It is a silicate of magnesia coloured green by iron, and possessing considerable lustre, though now but little valued in jewelry. It is found in rounded masses embedded in the cavities of basalt and lava : characters bearing some resemblance to those of the Callaina, as "sticking upon rocks, and protuberant like an eye." The difficulty remarked upon by Pliny in the polishing of the stone, which does *not* exist in the case of the Turquois, applies strongly to every species of the Peridot. "Augites" (Eye-stone), as the same writer observes, was considered by many to be only a synonym for the Callaina. It may be that this

remark has induced our mineralogists to give the name of
Augite to a mineral generally found in company with the
Olivine in its basaltic matrix; being nearly identical with
the latter in chemical composition, but black, and useless
for the jeweller's art.

The Olivine is the only precious stone of whose existence beyond the limits of our world we have proof positive,
for grains of it have been discovered filling up cavities in
aërolites, from whatever unknown regions in space those
wondrous visitants may have emanated.

CAMAHUTUM: *Camaut: Cameo.*

OF this word innumerable etymologies have been proposed, all equally unsatisfactory, as Menage long ago observed. The list begins with De Boot's simple and obvious derivation from Pliny's *Cyamea*, and goes on to the most far-fetched of all, Lessing's, who makes it the corruption of *Gemma onychina*, taking for his starting-point the Gothic form *Gamahuia*, to be found in Agricola, and explained by him as " gammen-hu," *bacon-stone,** a popular name expressive of the fatty streakiness of its composition.†

But the only safe mode for tracing out the true source of the word, so diversely written at various times, is to examine in what form it makes its first appearance in any European language. Of such, the very earliest example known to me occurs in the list of gems collected by our Henry III. for the embellishment of his projected shrine of Edward the Confessor, amongst which are described above eighty camei, of which fifty-five are particularised as "large." (This list is to be found in the Patent Rolls of his 51st year.) The word is spelt at length *chamahutum*, *camahutum*, and abbreviated *chamah'*, and *camah'*. His con-

---

* A notion arising from the use of *speckstein*, steatite, for their smaller works, by the German carvers of the fifteenth century.

† The list of etymologies winds up with the novel and ingenious one (the latest propounded) of Von Hammer's, from the Arabic of the same sound, meaning either a "flower," or the "top of the camel's hump."

temporary Matthew Paris Latinises it into quite the modern form (Vita Leofrici), "lapidibus nobilibus insculptis quos *cameos* vulgariter appellamus." Later, in the year 1341, we find the great French Cameo entered in the inventory of the jewels of the Ste. Chapelle as "unum pulcherrimum *camaut*." Now, the term is evidently of foreign origin, and in these early lists it seems to designate rather the species of the stone than the nature of the work upon it, which latter the writer in each case particularly describes. *Camaut*, therefore, may be safely put down as the original form of the word. Now *Camea* is said to be Hebrew or Arabic for a talisman, for which with either nation every engraved stone passed. It long retained that meaning : Gaffarel, in his 'Curiositez Inouyes' (1632), has, "Figures ou Images naturelles, *Gamahé* ou *Camayeu*, tiré par adventure du mot Hebreu, *Chemaia.*" In another place, however, he derives the word from the French *camaieu*, which he says means a figured Agate, evidently supposing the latter an indigenous word. This word *camea*, however spelt, has puzzled philologists, but the true connection of ideas appears to me to be this. Ben Mansur has (Div. ii. No. 2), "The *Camahem*, called by some the Ass-stone ; it is very hard, can only be pierced by the Diamond ; if broken, it separates into fine splinters. Rubbed down on a hard stone, it gives out a *red* colour. The finest sort is the Black-Red dug up in the district of Karak." Meaning unmistakeably our compact Hæmatite ; giving its distinctive characters, the iron-grey colour (that of an ass-hide), its hardness, splintery fracture, and red streak. Now this stone being frequently magnetic is the substance chosen, before all others, and from the very earliest times, to engrave talismans upon. Its use begins with the Babylonian Cylinders (most abundant in this), and ends through the Gnostic series with the Cufic stamps of the 8th and 9th centuries. Thus we can perceive

how *Camahem*, the common material for a talisman, came to
be identified with the talisman itself. Camillo, writing
at the close of the Middle Ages, certainly has the Camahem
in view when talking of the stone Kaman, and Kakaman, a
name he, with his favourite ostentation of his small knowledge of Greek, derives from Καῦμά, from its volcanic
origin, adding, " it is like the Onyx : " by which comparison
he doubtless refers to the *engraved* Cameo. These latter
works he actually cites by the name of ' Chamaini '.when
contrasting them with the similar reputed lusus naturæ in
stone.

I am tempted to think that *Camaut* came to be applied
to gems in relief, as being accounted talismans *par éminence*,
seeing how the Arabs have ever looked upon all ancient
bas-reliefs as magical ; and so the word travelled into
Europe in the vocabulary of the Crusaders in this restricted
sense. Intagli, on the other hand, the latter continued to
use whenever they could procure them for personal seals,
secreta, and distinguished such by the name Sigilla (the
diminutive of Signum, in its sense of signet), in popular
parlance termed ' *Pierres d'Israel*,' as the supposed works
of the ancient sages of that race.

Important Camei are not unfrequent which have been
converted into Gnostic amulets by the engraving of their
customary types and formulæ upon the reverse. I have
noticed examples in the Royal and in the Marlborough
Collections. The latter possesses a fine Cameo bust of
Commodus : on the reverse has been added a figure of the
Abraxas-god surrounded by the legend :

ΑΡΔΟΤ ΓΕΝΝΑΙ ΟΔΕΜΜΕΝΑΙ ΒΑΣΙΛΙΣΚΩΣ.

This superstitious ancient practice supports the derivation,
now the most popular, of " Cameo," from the supposed
Hebrew or Arabic word " camea," *a spell.* Such additions

(and they are not unfrequent) testify to the feeling with which similar glyptic works were regarded in the ages when the night of credulous barbarism was fast spreading over the Roman world.

The original "Camahut" was metamorphosed in mediæval mouths into the ugly Teutonic *Gammahuia*, and so accepted as a translation of "Speckstein" into the soft Italian *Cameo* (already used by Cellini, with whom, however, it signifies the stone, Onyx, not the style of engraving); and lastly into the fantastic French *Camayieu*, the form under which the word first appeared in English. But, very early, "Cameo" came to signify the work itself, not the material, for De Boot (who may be said to belong to the Cinque-dento period, for he wrote in 1600) notes that the Italians termed the Onyx, when engraved with figures in relief, "Cameo" or "Camahuia," as if the latter were a different substance. •

SHELL-CAMEO WORK.

THE antique productions of this branch of the Glyptic art will be found treated of under their peculiar and precious material the *Sardonyx;* but this appears the most fitting place to notice an ingenious application of the same art giving value to another material in itself the very synonym for worthlessness. This is the working in shell-cameo; the invention, it would seem, of the acute Italians at the very dawn of the Revival (1450), when the insuperable difficulty of procuring the Sardonyx to subserve their first attempts at imitating the ancients in cameo-engraving obliged them to look about for other substances affording the same capabilities in the way of contrasted colours for the exhibition of the relief. This vehicle for their skill they discovered in the larger shells of their own seas, presenting

thin strata of opaque white upon a buff ground, and
exactly imitating the Agate-onyx of two layers only. In
this the artists of the early Cinque-cento, even the very ablest
of their school, produced the most elaborate and carefully-
finished designs, perhaps encouraged by the extreme facility
of the mechanical part of the work, which is performed
entirely with the graver as in executing woodcuts. Of
the productions of this date the most noted are the bracelets
of Diana de Poitiers (Bibliothèque Impériale), consisting
each of seven oval plaques linked together, bearing sever-
ally one or more animals pertaining to the chase; a lion, a
wolf, a stag, a wild bull, a group of horses; in accordance
with the absorbing passion of the fair owner. The settings
are enamelled with the well-known cypher of her royal
lover; and the work is ascribed, but only upon plausible
conjecture, to the contemporary and famous engraver to the
Mint, Matteo del Nassaro. The Marlborough Cabinet
includes, amongst its other rarities, some wonderful specimens of the same class, without a name, indeed, but
infinitely surpassing the last mentioned in artistic merit.
The first is a battle-piece, full of figures grouped with a
spirit and an intricacy reminding one of Albert Durer's
celebrated battle-cut in the 'Triumph of Maximilian.' The
second, on a somewhat larger scale, gives us to the life the
Indian Triumph of Bacchus, where the god returns enthroned
upon his elephant, escorted by a countless train of bacchantes
and satyrs. The third is the very earliest memorial of
this invention to be seen anywhere: it is the busts side by
side of the Three Kings, each exhibiting his national
complexion in the natural colours of the shell, of white,
tawny, and black, marking the personifications of Europe,
Asia, and Africa. The stiff manner retains much of the
Gothic, and indicates the date of this cameo as not later
than the middle of the Quatro-cento period.

Such early shell-camei were doubtless abundant in their time, but have been destroyed on the subsequent revolutions of the mode, for the sake of the intrinsic value of their mountings; the fragility of their substance not allowing them to brave all such accidents with impunity, like their prototypes in "hard stones." After the above-mentioned period, this art seems to have become extinct; the material was considered too worthless to be redeemed by any amount of taste it might embody. But in our own times it has taken a fresh and a prodigious development, and is indeed the sole, though bastard, scion of the Glyptic art that yet flourishes in Italy. This revival is due to the introduction of the West Indian conch into the cameo-cutter's atelier; its numerous and richly-coloured layers of varying depths emulate and often excel those of the choicest pieces of Sardonyx possessed by the ancients. The seat of the art is at Rome and at Naples. At the latter place, from their lower reputation, the works are purchaseable at much lower prices: the subjects are perpetual repetitions of noted bas-reliefs, pictures, or gems. An interesting application of the invention is to the perpetuation of portraits either modelled from the life in wax and thence copied on the shell, or, now, done with great fidelity after photographs transmitted from abroad. Saolini bears at present the highest reputation in this line, and has of late, as I am informed, had the ambition to advance to the nobler field of engraving in the true and proper material of his profession. Cameo-engraving in shell has also, within the last quarter of a century, been transplanted to Paris, in whose kindly soil it flourishes as vigorously as when originally cultivated there three centuries ago under the patronage of François I. and Henri II. The French workers—more ambitious than the Italian, the nation of copyists — scorning slavish and unvarying imitation of

others, strike out original designs inspired by the taste
peculiarly their own. Nothing can be more charming than
the groups of Cupids engaged in infinitely diversified but
always graceful occupations which the Parisian artist
delights to imagine, and nothing also can be more appropriate for the purpose to which these camei are usually
applied.

CERAUNIA.

By this name two very different things are designated, according as the ideas expressed are a resemblance in the stone to the brightness and flashing of lightning, or its being itself the actual missile shot from the clouds. The first part of Pliny's description (52) applies clearly to either the bright Blue Beryl, or to the paler and more lustrous kind of Sapphires. " Amongst colourless (candidæ) gems is that called *Ceraunia*, which snatches its lustre from the stars (alluding to its celestial origin). Of itself it is like crystal, but its lustre is sky-blue; it is found in Carmania. Zenothemis allows that it is colourless, but says it contains a sparkling star; there are also Cerauniæ that are dead, yet if steeped for a few days in nitre and vinegar they conceive the same kind of star, but lose it again after an equal number of months." This would seem to indicate an Asteria or Star-Sapphire; in fact, the Ceraunia is termed an inferior kind of *Astrion*. Later, however, the name seems to have been applied to the Ruby, to distinguish it from the other numerous varieties of the Carbunculus, for Tertullian (' De Anima,' ix.) uses the simile: " In the gem *Ceraunia* the substance is not necessarily fiery because they sparkle with a ruddy blush (rutilato rubore)." Isidorus also makes two sorts: one like Crystal, but blending an azure with its red, found in Carmania; the other from Lusitania, like the Pyrope, but setting the fire at defiance,—the last character a proof that he is speaking of the Ruby.

In these gems, therefore, *Ceraunia* or *Ceraunias* was no more than an epithet, " Lightning-like," applied to any especially lustrous stone of the Corundum class.

But the second part of Pliny's notice introduces a very different and most curious idea.   Isidorus indeed pretends that the gem falls from the sky in thunderstorms, but this is merely a theory coined by himself to explain the name, for he adds κεραυνὸς in Greek signifies lightning.   But Sotacus, as quoted here by Pliny, " makes two other kinds of the Ceraunia (besides those first noticed), the black and the red, resembling *axes* in shape. Of these, such as are black and round are holy things, cities and fleets can be captured by their means, and they are called *Bœtyli*, but those of an elongated form *Ceraunæ*. Others make out a third sort, greatly sought after by the Magi, inasmuch as it is only found in places struck by lightning."

This last remark seems to have been the origin of the etymology of Isidorus referring to gems of a totally different character.

The comparison of the second and third class of Ceraunæ to *axes*, at first sight so inexplicable, becomes quite intelligible when we recollect that the popular German name to this day for the stone-axes of the primeval Celts is " Donnerkeil," or thunder-bolt.   Such the country-folks still believe fall from the sky in storms, and are to be dug up wherever the lightning has penetrated the earth.   Thor, the Teutonic Jove, is represented armed with a huge stone-hammer, in the place of the classical winged thunderbolt (which, by the way, is always pictured as a double-pointed cone, but equipped with wings).   As these primeval weapons are often exhumed in the neighbourhood of old Teutonic temples, it is probable they continued to be used as sacri-

ficial implements (in memory of ancient customs) long
after they had been superseded by metal in common life.
We know that the Romans in ratifying a treaty killed the
victim with such an instrument, "silice percussit;" (whence
the proverb "inter sacrum et silicem"), which could not
have been a stone casually picked up, but one sharpened
and fitted to a handle, so as to be capable of dealing a
mortal stroke at one blow. Charun, the Etruscan god of
Death, wields, in the tomb paintings, as his peculiar
weapon, an enormous mallet.

These considerations explain why Sotacus terms such
axe-like stones "holy." They are still highly reverenced
by the German boors, and considered a specific against all
diseases in cattle: hence Kerner, who has figured many of
them (in his little treaties on amulets), had the greatest
difficulty in purchasing specimens from the lucky finders.*
The distinction Sotacus makes as to the difference in their
shapes agrees wonderfully well with that to be noticed in
existing examples. His *Bœtyli*, black and round, are the
hammers perforated to receive a handle (*Beetle* is Saxon for
mallet): the more elongated *Ceraunia* seem analogous to ·
the stone celts of an earlier manufacture, and more plentifully discovered.

These stone axes puzzled the learned of the 16th century
to account for their origin fully as much as the ruder forms
now exhumed at St. Acheul do our geologists. Gesner
(1562) adduces the testimony of eye-witnesses to the fact
that such had been found in a windmill immediately after
it had been struck by lightning, and similarly under an
oak, and other places. He mentions the popular belief that
the "Strahl-pfeilen" or "Strahl-hammern" (lightning-

* Similarly the Italian peasants of the Marche highly prize the flint
arrowheads they pick up in the *terra mara* of the old Gallic sites, as
certain amulets against danger of lightning.

arrows or hammers) penetrated into the earth to the depth
of nine fathoms, and then rose up towards the surface at
the rate of one fathom in a year. De Boot quotes many
similar instances of their discovery under the same circumstances, but, unable to resist the evidence of human workmanship which they exhibit, ingeniously conjectures that
they had been originally iron tools, but converted into
stone by long continuance in the earth!

The *Iris*, mentioned by Pliny as next in estimation to
the Ceraunia in the opinion of the Magi, was nothing more
than a piece of Rock-crystal which had accidentally assumed the form of a prism, so that within a room it was
capable of forming the prismatic spectrum upon the wall.
A most beautiful phenomenon truly, and well adapted to
excite a belief in the supernatural virtues of the medium in
the minds of people ignorant that the effect was due merely
to the form.   It was remarked as strange, that when placed
in the sun the arrangement of its angles did not produce
the rainbow colours, but only when under the shade of a
roof. Pliny observes that the stone itself did not contain
the colours, but that they were elicited by the reflection of
the wall. In all other respects it resembled common Rock-
crystal : it was found in an island in the Red Sea, sixty
miles off Berenice.

There was another gem resembling this in appearance,
but not in effect, and therefore called *Leros* (empty), which
was a Crystal crossed by a white and black cloud.

The gem now known as the *Iris* is Rock-crystal containing many natural flaws, which by the refraction of light
produce the most beautiful play of colours internally, resembling sheets of flame.   The Iris can be artificially
created by flawing the Crystal with a sharp blow of a
mallet, or by throwing it into boiling water.

Besides the Ceraunia, Pliny (65) enumerates other stones

that fall from heaven amidst thunderstorms and rain, such as the Ombria, called by some the Notia (Rain, or Scirocco-stone), and the Brontia (thunder-stone), all supposed to possess the same virtues. It was also pretended that if the latter were laid upon an altar, the offerings could not be consumed so long as it remained there. Another version, "if we have faith sufficient," says Pliny, was that the Brontia got into the heads of tortoises after thunderstorms, and was to be found in their brain. This consequently possessed the virtue of extinguishing all fires caused by lightning.

The fossil Belemnite, popularly called Thunderbolt in England, was a few years ago (in my recollection) universally believed to fall from the sky under the same circumstances as the Ceraunia; perhaps, indeed, the Bætyli of Sotacus, "black and round," may have signified this petrifaction, not the artificially-made celts. That these fossils were counted amongst the precious stones of primitive Man is not a mere matter of conjecture. In the very early tumulus on Langton Wold, opened in June, 1865, amongst the other ornaments interred therein with a female, shells, jet-beads, bronze pins, &c., was found "part of a Belemnite ground and polished to the shape of a roller."

(G) G

CHALCEDONIUS: *Calcedony.*

THE Chalcedonius of Pliny was an inferior species of the Smaragdus, so called from being found in the copper-mines near Chalcedon. The supply of these particular stones had failed before his times, the mines being no longer worked; but he describes them as having been small and brittle, changing their colour when moved about, like the green feathers in the necks of peacocks and pigeons. But from the remark of Theophrastus (25), that they were seldom found large enough for ring-stones, but were used for soldering gold, and answered that purpose quite as well as Chrysocolla, it is evident that they were only crystals of transparent Chrysocolla, still popularly termed the Copper-emerald, and which to the eye have much the appearance of the true Peruvian gem.

It is difficult to trace the steps by which this name has been transferred from a substance of a brilliant green colour to one so totally distinct in all its characters as is our Calcedony (White Carnelian), a semi-opaque quartz of a milky tinge, and answering exactly to Pliny's Leucachates. The transition would appear to have been thus effected. The epithets Calchedonius and Carchedonius, so readily confounded together in the uncertain orthography of MSS., became inextricably intermingled in the ideas of the mediæval mineralogists. Now Pliny's Carchedonius, so called as being brought from Carthage (Καρχήδων), was a Carbuncle, showing as it were stars within its substance, and of sufficient size for cups to be cut out of it. Epiphanius and the Vulgate render Καλκηδών (the third stone in the foundations of the New Jerusalem), by " Smaragdus," but the Authorized Version translates it " Carbuncle." Marbodus adopts both renderings, for in his ' Lapidarium ' he places the Chalcedon between the Emerald and the Beryl in colour, only paler,

" Chalcedon lapis est hebeti pallore refulgens ; "

whilst in his ' Prosa ' he makes it a Carbuncle, lustrous by night alone:

" Pallensque Chalcedonius ignis habet effigiem."

Now, Pliny does in fact mention one kind of Jaspis as found at Chalcedon, but all he says of it is that it is cloudy (turbida). Hence it is probable that the " pale " colour, and slight " opal essence," somewhat resembling that of the Fire-opal, which distinguish that common European stone now known as Calcedony, gave it in the Middle Ages an undisputed title to that name; a transfer to which people were also led by the definition in the ' Prosa ' of Marbodus, their great text-book on the nature of precious stones. It may, moreover, be conjectured that Pliny's mentioning a " Chalcedonian Jaspis " also contributed to bring about this interchange of names.

The modern Calcedony is a semi-transparent white quartz (Quartz-Agathe-Calcedoine of Haüy), slightly finged with yellow or blue. Some of the best specimens of the former kind can hardly be distinguished from the poorer Hungarian Opal, for they exhibit precisely the same milkiness of tint; but their lustre is flamy, not iridescent, and contains no admixture of *green*. This is the *Girasol* of writers of the last century, but must not be confounded with the Girasol, or Chàtoyant, Sapphire, Pliny's Asteria. This

variety was evidently classed by Pliny amongst the commoner European Opals, which he distinguishes from the
incredibly precious and true Indian species. The kind
tinged with pale blue, now called the *Sapphirine* on account of
its colour, seems to have been the Jaspis Aërizusa, *i.e.* of a
colour *tending* towards the cerulean : this, adds Pliny, the
Greeks term *Boria*, from its resemblance to the sky on an
autumnal morning. An apt comparison, which exactly
represents the paley-blue tint of a good Sapphirine. Mar-
cellus Empiricus designates it the " cerulean Scythian
Jaspis " in his prescription for the making of an amulet
against the pleurisy. A third variety, found in large
masses, was turbid, very opaque, and soapy, sometimes
even of a dirty brown.* The two last species were abundantly employed by the ancients of every period, but more
especially by the Asiatics ; the Babylonian cylinders are
frequently made of grey or brown Calcedony, and it is the
material almost exclusively used for the large conical seals
of the Sassanians. The most beautiful Persian cylinder
known is in Sapphirine. It bears the usual type of the king
fighting against a lion (the evil genius), a design which
proves it to have been the royal signet that graced the
wrist of some Artaxerxes or Darius of the later times of
the Achæmenian dynasty. Scarabæi of Etruscan work, as
well as good Greek and Roman intagli, frequently occur
in this material, but engraved on the Sapphirine in preference to any other sort; and justly so, for it is an extremely pretty substance, often approaching to the pale
Sapphire in colour, though devoid of its brilliancy. Upon

* This again may be the *chernites* of Theophrastus (6) exactly resembling ivory " in a coffin of which Darius, according to report, is
buried." A beautiful opaque Calcedony (often forming one stratum
in the true Indian Sardonyx), has the modern name of *Cacholong*, said
to mean in the Cossack tongue " pebble of the river Cache," from whose
bed the finest specimens are brought to Odessa.

a large stone of this sort is engraved the finest Greek intaglio that is possessed by our National Collection : the subject is Victory crowning a trophy, in design identical with the same type upon the medallion of Agathocles, to whose date it indubitably belongs, and, in truth, may on very good grounds claim the honour of having been the royal signet. In form it is a *scarabeoid*, a modification of the scarabeus much in favour with the Asiatic Greeks. This magnificent historical gem was lately found at Syracuse. The Girasol is sometimes improperly called the Opaline, so that antique works in it are often described as executed in real Opal, a stone in which (with perhaps a solitary exception) they never occur.

Busts and heads, in full and in bas-relief, and of considerable volume, were executed by the Romans in Calcedony. Under the Empire, these carvings constituted the *phaleræ* so often mentioned as military distinctions. They had replaced at an early period the original gold bosses forming that decoration—Virgil's

" Phaleras Rhamnetis et aurea bullis—Cingula ; "—

for M. Cœlius, who perished with Quintilius Varus, is sculptured on his cenotaph (Mayence) with all the rewards earned during a long military career, decorating his effigy ; torques, armillæ, and on his breastplate the phaleræ—a Gorgoneion and several Bacchic masks evidently akin to those about to be described. Most of the existing specimens will be found perforated with holes placed diagonally for the reception of the rods framing the several pieces together. The head of Horus is, under the Lower Empire, a special favourite. These large sculptured heads in Calcedony may indeed have also been female ornaments, and employed in a manner one would not have expected, to decorate the hair. This discovery was suggested to me by

the famous bust of Isis bearing the inscription in the
Ethiopic characters (Turin), whose hair, twisted into two
long ropes, is looped upon her breast by passing diagonally
through a mask evidently formed of a gem, and again
crossing below is confined by a ring, and the same a third
time. It is still the fashion with the Greek women to string
perforated coins upon their hair twisted into many strings.
The most important of these relics are the Marlborough
Medusa in pure lustrous Calcedony, in the grandest style,
originally the central " decus et tutamen " of an emperor's
breastplate, and the bust of Matidia, supported upon a pea-
cock (typifying her deification), and three inches high; a
work of the highest merit in point of art, though the material
is brownish and turbid. Such monuments afford another
proof that this substance was included amongst Pliny's inferior Jaspis-gems : perhaps his Capnias, " stained as it were
with smoke," for he goes on to mention a statuette of Nero
in armour, 15 inches high, formed out of one piece ; and in
no other substance do antique works in full relief occur of
such magnitude as in Calcedony. Another very grand
work, in three-quarters relief and of considerable dimensions, is a bust of Tiberius, in a style of the utmost perfection, executed in an opaque species much like ivory; perhaps a real Cacholong or impure Opal * (Vienna). Amongst
the Marlborough gems, No. 384, is a Cupid's Head in the
most lustrous Girasol imaginable. It seems to have been

---

* In the Townley Coll., Brit. Mus., is a child's head, of considerable
size, admirably carved out of this substance : the resemblance to ivory
is most striking. It appears to be the stone described by Pliny (54) as
the "Arabica, exactly like ivory, and would pass for it did not its
hardness betray it. It is thought to protect the wearers against pains
in the sinews (rheumatism)." This appears to be the same as the
Thelycardios (68), " so charming by its opaque white colour the Persians
(its native locality) that they give it the name of *Mulc* (the King.)"
(See page 84, for " Chernites.")

originally one of those boy's (Horus) heads in a rude style, so much in vogue under the Lower Empire for *phaleræ*, for the usual perforation through its thickness is still perceptible. Some good hand of the Cinque-cento, however, tempted by the unusual beauty of the Calcedony, has entirely re-worked and improved the relievo, and by the addition of a pair of folded wings and an elegant open-work border in gold, has converted it, new christened into a cherub, into the most graceful pendant that ever adorned the neck of a Medicean princess.

That our two commonest species of the stone were the Leucachates and the Cerachates, White and Wax Agates of the ancients, can hardly admit of a doubt. Dionysius describes the Choaspes as carrying down in his torrents the beautiful *Achates* torn by the wintry rains from the mountains, and experience attests the almost exclusive use of the Calcedony for signets by the Persians of his age.

In consequence of the abundance of this stone in Europe, (almost the sole fitting material for the Glyptic Art produced by this region of as fine a quality as by the East), a larger number of Renaissance and modern works in relief will be found in Calcedony than in any other stone; whereas genuine antique Camei in this single-coloured material are much rarer than upon the stratified species of the Quartz family. De Boot has (ii. 88), "In our times (the close of the Cinque-cento period) cups, figures of princes, chaplets of beads, and an infinity of other things are made out of Calcedony. Its chief use, however, is for seals, because the wax (*i. e.* the soft wax then used) does not stick to it." To the correctness of the last remark ample testimony is borne by the large number extant of clever intagli of this date in common white Calcedony, a stone never touched by a *good* master in antique times.

CHRYSOCOLLA: *Native Verdigris.*

THE mineral which still retains its ancient title is a carbonate of copper, sometimes opaque, sometimes translucent,
and hard as quartz. The ancients obtained the best quality
from the copper-mines; an inferior, from those of gold.
It was also prepared artificially by steeping the ore during
the entire winter, and evaporating the liquid by the
summer heat (Plin. xxxiii. 26). The best came from
Armenia, the second quality from Macedonia, but the chief
supply was drawn from Spain. It was used in painting,
and is reckoned amongst those " florid " colours which on
account of their intrinsic value were furnished by the employer, not by the artist. It entered into the list of medicines, as a caustic application for wounds, ulcers, sore
eyes; and as an emetic, a very effectual one certainly, but
highly dangerous.

But its principal value was as a solder for gold (whence
its name, signifying *Gold-cement*), for which purpose it had
been employed from the earliest times, being so mentioned
by Theophrastus (26). For this purpose it was steeped in
a little boy's urine, together with verdigris and nitrum,
and then rubbed down in a copper mortar with a copper
pestle. This solder the Romans called *Santerna.** Gold

---

* Either a Punic or Spanish word, as are most of the Roman terms
used in metallurgy; unless from its termination in *rna*, the Etruscan
language, delighting in such endings, has a better claim to it : and

containing an alloy of silver could best be soldered with this, and was all the brighter at the junction; but if alloyed with copper, the gold shrunk up, grew dim, and refused to unite. For such a quality, by adding a seventh part of silver and a little gold to the above ingredients, and grinding them up together, a good solder was produced.

The ancient jewellers with this mixture turned out much better work than the modern : the soldering in their fine-gold jewels cannot be detected even by breathing upon it. Castellani, that most skilful reviver of the Etruscan style of jewelry, is forced to confess that the secret of their soldering (more particularly in the fine granulated work) is all but totally unattainable. That now used—an alloy of gold, silver, and arsenic—becomes white submitted to this test. The old recipe appears to have been preserved by tradition throughout the Middle Ages, for Cellini gives the ingredients of his gold-solder as : Native Verdigris (another name for Chrysocolla) 6 parts, Salammoniac 1, Borax 1, to be ground in water, and applied to the edges required to be joined with a brush, before going to the fire.

Nero, as the patron of the Green Faction, in one of his fits of extravagance, caused the Circus to be strewed with the powder of this costly mineral, instead of the ordinary sand, on the day when he figured there as a charioteer in a livery of the same dye.

the Etruscans were for many ages the goldsmiths of Italy. Pliny's *nitrum* was our *natron*, native carbonate of soda, the ancient substitute for soap. It is still used as a flux for gold-ore.

CHRYSOPRASIUS: Χρυσόπρασος.

AFTER noticing the existence of several kinds of green-coloured gems not considered worth a particular description, Pliny defines the Prasius, and then proceeds to state that there was a superior variety of the same, the *Chrysoprasius*, also of the same leek-green tint, but " receding somewhat from the Peridot's into a golden." It occurred of such dimensions that the oval cups termed *cymbia* (the *gondoles* of the French) were formed out of it. This last circumstance proves of itself that the stone would now be referred to the class of green Jaspers, in which several of these antique vases are still remaining, and have been fully noticed under Smaragdus. His *Chrysoprasius*, however, as distinguished from the above, is the third kind of his Beryl, approximating to the Chrysoberyllus, but still paler or yellower, and considered by some as belonging to a distinct species. This latter supposition was to all appearance correct, and the stone in question the Indian Chrysolite. Most certainly it was not our Chrysoprase, Silica coloured a beautiful apple-green by oxide of Nickel, slightly translucent, and of uncommon hardness. Had the ancients known the latter gem, they must have classed it amongst the more opaque Smaragdi: it is more probable they were unacquainted with it, for its only source at present is the mine of Kosemütz,* in Silesia. Although of much the same com-

* Where it occurs in veins traversing Serpentine.

position as the Prase, it differs greatly from that stone in outward appearance, by reason of its opacity and much more agreeable and more uniform tint of green, besides its greatly superior hardness.

Although antique works do not occur in our Chrysoprase —a necessary consequence from the fact that the only locality producing it lay without the limits of Roman enterprise—yet a mineral much resembling it, though with somewhat of a bluish cast, is said to be found sometimes set in old Egyptian jewelry alternately with bits of Lapis-lazuli. Intagli also are known upon this same substance, which has, on account of its colour and slight translucency, been mistaken occasionally for the Turquois. One in my possession bears a most singular type, Barbelo, the androgynous "Mother of Creation" of the Phoenicians, with the appropriate attributes of the snail (salacity) and the butterfly (life). The substance may be defined as a blue Plasma ; it contains the peculiar black specks that characterise the species. This antique kind may with some probability be considered the cerulean Jaspis brought from the Thermodon.

Some such stone as this may have been the Omphax, reckoned by Theophrastus (30) amongst the commoner ring-stones. The word, primarily signifying an unripe grape, could only have applied to a gem equally opaque and green, with a certain bluish bloom upon it. It is strange, indeed, that no Greek intagli are now extant in any gem at all answering to this description, for Theophrastus puts it in the same list as the Carbuncle, Sard, and Amethyst (Garnet), whence one would conclude it to have then been equally in general use. Furlanus understands by it an inferior sort of Beryl, De Laet says "de quo nihil certi possum affirmare." A recent writer is certain that our pale, uniformly-tinted Prase is signified : but the translucency of the last species seems to me to controvert his attribution. There can, however, be no doubt some variety of green Jasper was the stone in question.

Epiphanius mentions under " Chrysolithus " a variety, the *Chrysopastus*, or gold-spotted. This afterwards came to be confounded with the Chrysoprasius, for Marbodus makes two kinds of the latter—one that of Pliny, named from its resemblance to the leek-leaf, the other purple with golden spots. After these he makes a third, of a similarly-compounded name, the Chrysoprasion, which blazes by night, but resembles a piece of gold in the daytime, applying to it what Pliny (52) says of the Chrysolampis.

At the beginning of our century this beautiful stone was much in fashion, both set round with small brilliants for brooches and rings, and also singly for necklets and bracelets. The paler pieces were coated on the back with verdigris, to heighten their tint. No other coloured gem so well becomes a blonde complexion. But it loses its beauty, fading away to a mere white Calcedony if long kept in a warm and dry place, or if constantly exposed to the light, still more rapidly if to the direct radiance of the sun.

CHRYSOLITHUS: Χρυσόλιθος: *Oriental Topaz, and the Jacinth.*

THE Chrysolithus of Pliny (42), or at least his best sort, the Indian, was the gem now commonly but improperly styled the Oriental Topaz, a yellow variety of the Sapphire, and of equal hardness and rarity. The ancients obtained it from Ethiopia (a vague term for the remote East), together with the Hyacinthus (Sapphire): a natural companionship, both being Corundum but differently coloured—the blue and yellow *Jacut* of the Persians. The description, "transparent, with golden lustre," applies to no other gem so exactly as to this. Such is its brilliancy that when De Boot wrote it was considered superior to the White Sapphire for imitating the Diamond, after the colour had been extracted by heat.

In the first class were placed the Indian, and those brought from Tibara, if not cloudy (turbidæ). The test of their quality was that their intense yellow should make gold compared to it look as pale as silver itself. This golden lustre is a conspicuous quality in the Oriental Topaz, the Brazilian, on the contrary, being betrayed by a vinous tinge. The Arabian Chrysolithi were most probably no other than the modern Jacinths, for Pliny's account of them applies exactly to the latter gem : " They are in least esteem of the whole class, being turbid and of different shades; and even when limpid their lustre is marred by a cloud of spots, as if they were filled up with their own dust" (scobe); an evident allusion to that porousness or granular and even bubbly texture so conspicuous in the Jacinth. Besides, a gem so much in fashion with the ancients as our Jacinth was could not have been omitted from Pliny's list, and here alone is a description to be found at all applicable to it (LYNCURIUM).* The same gem seems also intended by his Melichrysus, "transparent like pure honey shining through gold," an Indian stone, hard yet brittle. Also to the same family belongs the Xanthus (orange), "a vulgar gem in that country" (India). So unchangeable is Indian fashion, that it in our day may be adduced in support of this attribution: no stone is more common in Hindoo jewelry than the Jacinth.

On the other hand, his Pontic Chryselectri, "amber-coloured, and recognisable by their lesser weight† (the Oriental Topaz being the heaviest in our scale), some of them hard and reddish, others soft and cloudy," were no more than, the former, Cinnamon-stones, the latter, Rock-crystals, variously tinged with yellow, orange, or brown (Rauch-Topaz — Cairn-gorum). Their exact nature is settled by the fact quoted from Bocchus as to the magnitude of one, found in the Crystal-mines in Spain, which weighed twelve pounds; a sufficient proof that he is speaking here of those smoked Crystals, often improperly called European Topazes, which sometimes attain an incredible size: although we must accept with certain reservations De

---

* His remark that "they have now gone out of fashion for signet-stones" (tametsi exiere jam de gemmarum usu) supports this view, late Roman work never occurring on the Jacinth, all intagli upon it being either Greek or certainly anterior to Pliny's times.

† "Ponticas deprehendit levitas," a remark indicating that the ancients applied in their investigations connected with this science the principle of specific gravity, and perhaps even the hydrostatic balance of Archimedes.

Boot's record of a Bohemian Topaz (presented to his patron Rudolf II.), two ells long by one and a half broad.* Yet an elegant shell-formed cup (Renaissance) in this material, large enough to contain a half-pint, has come under my own notice.

By a singular coincidence, Pliny actually designates one variety by the Greek equivalent for the German name, *Capnias,* for *Rauch-Topaz.* Again, there was the Leuco-chrysus, a variety produced by an opaque white vein crossing its middle : a formation which would now cause such a stone to be referred to the class of Agates. The Chrysolithus was at that time counterfeited so exactly in paste that the deception could only be detected by the touch, to which the vitreous body felt warmer than the genuine : a proof, this, of the value fine-coloured gems of the class commanded in the Roman market.

Chrysolithi were, when of the first quality, set *à jour* (funda perspicua), almost the sole exception to the then uni-versal custom of backing the stone with gold : the inferior were foiled with *aurichalcum,* a red foil of gold, much alloyed with copper. Transparent gems, when extracted from the remains of their original antique rings, are frequently found backed by a leaf of *red gold,* of quite a different standard to the pure metal used in the jewelry of the same period. Pliny also mentions the backing of Carbuncles with a silver foil, a method still in use, and the most advantageous when the stone is of fine quality. The object in using *coloured* foils is merely to deceive, and to impose upon the inexperienced by thus imparting to an inferior gem the finest colour belonging to its own class.

Chrysolithi, whatever they may have been, were in high esteem with the Romans. They are ranked with Emeralds

---

* Meaning the Vienna ell of 27 inches.

by Propertius (II. xvi.) amongst the bribes by which the
" wealthy prætor from Illyria," gorged with the plunder
of his province, had seduced the poet's Cynthia away from
him—

> " quoscunque Smaragdos
> Quosque dedit flavo lumine Chrysolithos."

This " yellow lustre " of the Crysolithus, coupled with
its value, points out the Oriental Topaz as the gem here
meant. The Jacinth was much too common a thing then
to be mentioned in the same breath as the Emerald, the
most precious gem in the Roman jeweller's list.

The Oriental Topaz was formerly much esteemed, and
that in comparatively recent times. De Boot puts the
value at 2 thalers for the first carat; and after, as the
weight squared : Dutens, in the last century, at a third
higher than the Sapphire. The finest on record is that
seen by Tavernier (1665) in the treasury of Aurungzeb,
weighing 157¾ carats, and but recently purchased (at Goa)
by that emperor for a sum equal to 18,000*l.** At present,
their value is little more than nominal, because in wear
they are so easily confounded with the common stones of
the same colour.   .

This stone (the Oriental Topaz) was too valuable, and
perhaps too hard, for the ancient artists to attack ; no
genuine works of theirs are therefore to be met with in
it.† The only yellow gems employed by them were the

---

* " 181,000 rupees " which he makes = 271,500 livres, but in reality
it would amount to 18,100*l.* It was the only jewel Aurungzeb ever
wore when our traveller saw him. Its shape was an exact octagon,
with two rows of small facets along the top of the beasil.

† Cellini indeed mentions as one of the three finest antiques procured by him during his first visit to Rome (1524-27) a perfect Topaz
the size of a big hazel-nut, " una grossa nocciuola," engraved with a
head of Minerva ; but he probably means by Topaz no more than the
bright yellow crystal sometimes used by the Roman artists. Cellini

Jacinth, and occasionally the Cinnamon-stone, the latter frequently by the Sassanians. From some unexplained cause, the yellow Crystal, though equally abundant in nature with the purple (or Common Amethyst), was very seldom engraved upon by the ancients: genuine works in it being infinitely more unfrequent than in the Jacinth. The best examples known to myself are a head of Julia Titi (Rhodes), an exact replica in miniature of the famous Beryl by Evodus, and a large double-convex stone of great lustre covered with a Gnostic formula on both faces (Praun, now in the British Museum). Cinque-cento and recent engravers, however, have largely turned to account this pretty and easily-worked material, and therefore any intaglio of merit appearing in it must, *primâ facie,* be regarded with the utmost suspicion, nor be admitted as antique, until both the artistic part and the surface of the gem have satisfied the closest scrutiny.

First in the list of modern works may unquestionably be ranked No. 366 in the Arundel Gems. It is a bust of Philip II. of Spain upon a large oval stone of extraordinary purity and lustre, almost inclining one to believe it a yellow Corundum. The bust is in half-relief, and presents a highly-finished and life-like portrait of that monarch in early manhood: a *chef-d'œuvre* amongst the portrait-camei of the age, that epoch of perfection in this branch of the art, and worthy of the reputation of the court-engraver Jacopo da Trezzo, to whom it has from the circumstances every claim to be attributed.

Epiphanius says that the Chrysolithus is also called the Chrysophyllus: meaning, perhaps, either Pliny's Chrysop-

could have known nothing of what now exclusively usurps the name of Topaz, peculiar to Brazil, when a precious stone : for the Portuguese did not begin to settle in that country before 1531. Even at the end of that century the well-informed De Boot was certainly unacquainted with the Brazilian Topaz.

(G)   H

terus, or more probably his Chrysoberyllus; "a Beryl, paler green, which goes off into a golden lustre:" a description applying indeed better to our Indian Chrysolite than to any species of the true Beryl. For Epiphanius states that it is found in the deep quarry or shaft (φρέατι διπέτρῳ), near Babylon ; and Dion. Periegetes notices as the sole productions of that region its vast palm-groves, and the " Beryl, enclosed in the rocks of Serpentine." Powdered, and taken in drink, it was held a panacea for all complaints of the chest and bowels. The modern Chrysoberyl (p. 51) when opalescent becomes the Cymophane (wave-shower), only found in Siberia, and the most beautiful of stones: a wave of silver floats upon its golden green surface as it is turned towards the light, and plays therein with the vivacity of the Opal. It evidently was unknown until quite recently, and is as yet extremely rare in jewelry, and fetches a very high price. (See CAT'S-EYE.)

Pliny's Craterites, or " Stone of Strength," of a tinge between the Chrysolithus and Amber, and noted for its extremely hard nature, may with reason be supposed, on the ground of the latter quality, only an epithet of the Oriental Topaz.

It would appear that all this class of stones were esteemed in proportion to the depth of orange they possessed, for *yellow* was certainly not a colour admired by the Romans, except in *hair.* Pliny notices, speaking of *gold,* that its colour was not esteemed the finest of all either in *gems* or in other things. This was therefore the reason why their engravers neglected so completely the Yellow Rock-crystal, whilst the Purple was such a favourite with them.

The mineralogists of the Lower Empire certainly applied the name Chrysolithus in the sense it bears at present. This is deducible from the remark of Epiphanius above quoted : and placed beyond a doubt by the definition of it in Marbodus, as "a stone of a fiery colour mingled with sea-green;" an apt expression for the lustrous *Chrysoberyl.* It came from Ethiopia. If set in gold and strung upon hairs plucked out of an ass's tail, being worn as a bracelet on the left arm, it drove away demons, and dispelled the terrors of the night. De Boot notices that the jewellers of his times confined the name of Chrysolite to the ancient Topazius (our Peridot), and that it was then so common a stone that he does not take the trouble to fix its market-value. His contemporary, Shakespere, seems to have esteemed it more highly, for he uses as a synonym for incalculable price—

> "One entire and perfect Chrysolite."

At the same date, the Topaz meant a yellow stone, and was divided into the Oriental and the European.

Marbodus, even at his early epoch, distinguishes the yellow Corundum as the *Citrinus,* a designation, Citrino, still retained for it by the Italian jewellers, and much more appropriate than the erroneous and arbitrary distinction of Oriental Topaz.

His Citrinus, like the two other species of Hyacinthi, protected the wearers from danger in travelling, against pestilential air, and gave them favour with princes.

(For the *Chrysolite* of the moderns, see Topazius.)

u 2

CORALLIUM: Κουράλλιον: *Coral.*

THE Greek name for this zoophyte was derived, according to some (Plin. xxxii. 11), from the necessity of *cutting off* the plant while still living with a sharp steel (κουρὰ, shearing), for if touched by the human hand it instantly became petrified. It was considered to be a marine plant, green whilst growing, and producing white and soft berries, which, exposed to the air, hardened into the colour and the size of cornel-cherries; this latter tale bearing the true stamp of a Greek theory coined by their fancy to explain the origin of the beads first brought to them by their navigators from Massilia. It is briefly noticed by Theophrastus (38) as being a stone, blood-red in colour, but of a cylindrical form like the root of a plant, and growing in the sea; "the petrified Indian Reed" (the Arabian Coral) not being far removed from it in character. The ancient notion as to its vegetable nature rested not merely upon its shrub-like form, but also on the fact that its branches are clothed with a fleshy coating, soft whilst in the water, but drying up immediately upon extraction.

The Romans obtained their Coral from the Red Sea, but of too dark a tint to be in much request; from the Persian Gulf, where it was called *Lace;* but the best quality was fished up on the Gallic coast off the Stœchades isles (Hyères), and also off Lipari, and Trapani in Sicily. The Gauls before the subjugation of their country used it profusely in the decoration of their swords, shields, and helmets (this with

Amber being the sole ornament known to them in the way of jewels); but when Pliny wrote, the demand for it in India had become so great, that it was rarely to be seen in its native country, all that was in the least degree saleable being exported to the East; a notice this, by the way, that gives an insight into the vast commerce and the facility of communication existing at that time between Europe and India. Coral was as much sought after by the Indians as pearls were by the Romans, and fetched as high a price; the value of such things, as Pliny sagely remarks, being altogether arbitrary (ista, persuasione gentium constant). The beads (baccæ) were as highly prized by the men of the Indians, as the Indian pearls by the Roman ladies. They were worn as a specially sacred amulet, as well as for ornament by their " prophets and diviners (Brahmins); so that they delight in them both for their beauty and their supernatural virtues."

Every observation of Pliny's respecting the then trade with India is borne out by modern experience, Tavernier remarking that Coral was by far the most profitable article that could be taken out to that country. At present a sphere, sound, good-coloured, and weighing an ounce, will command a higher price than any precious stone, either in China or Japan, being in great demand for the apex to the mandarin's cap of office.

The Romans seem to have employed Coral merely as an amulet and in medicine : the little branches were tied round children's necks to keep off the Evil Eye ; and, powdered after calcination, it was taken in water for the stone and colic. Being of a cooling and astringent quality, it was an ingredient in salves for obstinate ulcers, for removing scars, and for complaints of the eyes. But Orpheus (570), who poetically fables it to have originated when the newly-severed Gorgon's head was laid down by Perseus on the

sea-weeds, which the issuing gore turned to stone, attributes
to the Coral a long list of the most transcendental virtues
bestowed upon it by Minerva.   It baffled all witchcraft,
counteracted poisons, protected from danger of tempests in
sea-voyages, and from robbers in land-journeys, and, mixed
in powder with the seed-corn, secured the growing crops
from damage of thunderstorms, blight, caterpillars, or
locusts.   These notions were embraced to their fullest
extent by the mediæval philosophers, as the section on
" Coral " in Marbodus proves.

On this account, Coral beads are often to be found in the
jewels of the Middle Ages in the same manner as the most
precious gems, and evidently equally valued.   The small
pointed branches, mounted with a ring at one end for suspension, so extensively manufactured at Naples, as every
tourist must have noticed, are still in request there as
amulets.   Ferdinand I. devoutly believed in their virtue,
and frequently pointed this minute defender against the
person whose malignant influence (*malocchio*) he suspected.
Phalli carved in Coral (the Neapolitan branch being merely
their modern and decorous representative) are worn yet
round the necks of the Roman females as a cure for
sterility.

It is not anywhere mentioned that the ancients ever
employed Coral for glyptic purposes, either in relief or
intaglio ; although a " Head of Chrysippus, in high relief,"
is quoted as antique in the Catalogue of the Orleans Cabinet.   The subject, however, as well as the material, makes
me suspect that this was rather a work of the Renaissance,
the engravers of which period frequently cut the thicker
pieces into Camei, but more especially into statuettes, skilfully availing themselves of the natural disposition of the
stem and its branches to form the body and limbs in the
attitudes required by the design.

In Pliny's alphabetical list of gems (56) occurs the *Corallis*, produced in India and Syene, and resembling vermilion in colour. This is an exact description of the antique Red Jasper, which certainly does not appear amongst any of the numerous varieties of the Jaspis, properly so called, described by him.

Pliny seems to have read the passage of Orpheus above quoted, for he has, under *Gorgonia*, " This stone is nothing more than Coral, the reason for the name being because it is transformed (from a soft plant) into the hardness of a stone. It appeases the angry seas; it is affirmed that it defends against thunderbolts and whirlwinds."

The same perverted taste that has in these days made rose-coloured Pearls far more valuable than those of the colour proper to the species, has pronounced that the same faint tint in Coral " is the only wear," and consequently the deep-crimson, hitherto the most esteemed, as sense dictates, has become a mere drug in the market. The pale pink, of the shade of the opening rose, has obtained, on the other hand, a value never dreamed of before in respect of this article. A parure in it at the Dublin Exhibition (1865) was priced at 1000*l.* Yet in most cases this precious tinge is artificial, obtained by baking the common sort until its red is blanched down to the required tone.

CRYSTALLUS: Κρύσταλλος and ῞Υαλος: *Rock-Crystal.*

THE Crystal (Pure Silica) is reckoned by Theophrastus (30) amongst the gems used in rings, where he speaks of it in company with the Amethyst, adding " both are transparent." Their joint mention shows his knowledge of the true nature of the latter, as being merely an accidentally coloured Rock-crystal. It is curious that, in spite of this inclusion of it in his list of ring-stones, intagli of the Greek, or indeed of the Roman period, upon Crystal, are unknown to collectors. Herodotus (iii. 24) mentions a stone by the name of ὕαλος, which, if there be any truth at the bottom of his story, must have been Rock-salt, judging from its magnitude and facility in working. " The corpse (of the defunct Ethiopian amongst the Macrobii), after having been dried, whether in the Egyptian or some other manner, is coated with plaster, and painted so as to imitate life, and then put inside of a hollow pillar of *hyalos,* which is dug up there in large quantities, and is easily worked ; and the corpse, being enclosed within the pillar, shows through it, without producing a stench or anything unpleasant, and wearing all the clothes he used in life."

Rock-crystal, however, was in enormous request amongst the luxurious Romans under the Empire for the purpose of making drinking-cups, valued as highly as the Murrhina, with which they are generally associated in the allusions of the satirists. An equal mania possessed the fashionable Romans for the acquisition of each sort, the Crystal being appropriated to iced, the other to boiling liquors. Thus Pliny instances a lady, and she not a wealthy one, who had then lately bought a Crystal *trulla* (flat bowl) for a sum equivalent to 1500*l.* of our money. Nero, to revenge himself upon mankind by preventing any one else from drinking out of the unique vessels, when informed that all was lost, dashed upon the floor and smashed two Crystal deep basins (scyphi) engraved with designs from Homer. Seneca tells a story (De Ira, iii. 40) that strongly illustrates their enormous value; how Augustus punished his friend Vedius Pollio, who had ordered a boy for having broken a crystal vase to be thrown into the pond where the lampreys were kept (to be eaten alive by them): the culprit, flying to the emperor's feet, petitioned for a less horrible death; Augustus, as a punishment commensurate with the cruelty of the sentence, ordered all Pollio's vases of the kind to be broken to pieces in his presence, and the pond to be filled up.

Some idea of the various employments of these two materials amongst the Romans, with whom they filled the place now occupied by Sèvres and Dresden porcelain, may be derived from Fauno's account of the articles found in the sarcophagus of the child Maria, the betrothed bride of the Emperor Honorius, discovered in making excavations at St. Peter's in 1544. "Within a silver box were vases and articles in Crystal, thirty in all, big and little; amongst which were two middling-sized cups, one round, the other oval, engraved with the most beautiful figures in *intaglio*. Also a snail-shell (or Nautilus) in Crystal, fitted up with fine gold as a lamp. The gold covered the orifice of the shell, leaving only an opening in the middle for pouring in the oil, by the side of which is fixed on a pivot a golden fly, moving to and fro for the purpose of covering the

opening. In the same way the nozzle and hole for the wick
are drawn out in a sharp projection with the greatest elegance, and so united with the Crystal that it appears all
one piece naturally. The cover, too, is equally well wrought.
The shape of the shell is like a large sea-shell, encompassed
all round with its points, which in this vessel are polished
and very smooth, the Crystal being so well worked. There
were also vases and different articles in Agate, with certain
little animals, eight in all. Amongst these, two very beautiful vases, one like those glass ampullæ made big and squat
for holding oil or such like liquids, so wrought, so elegant,
so thin, that it is a marvel to behold; the other, in the
shape of those brass ladles with long handles used in Rome
for baling water out of cisterns, and supposed to have been
a vessel used in sacrifice by the ancients."

These Crystal vases were either embellished with raised
mouldings and gadroons, much after the manner of the
modern cut-glass, or, if engraved with foliage and figures,
these were always in *intaglio*, the " crystallum impunctum "
of Apuleius. For to all ancient ornamentation in *relievo* a
contrast of colours was a necessity, to supply a ground for
the figures, which would have been lost upon a substratum
of one transparent colourless body. Another advantage
Pliny notices, drawn by the artists from this mode of treatment—the *incised* work enabled them to cut out or disguise the flaws and red patches too frequently disfiguring
the substance.

Many of these vessels were of extraordinary dimensions,
though Pliny shelters himself under the authority of Xenocrates for quoting one as large as an amphora (a cubic foot),
and another, brought from India, holding four sextarii (two
quarts). The crowning glory of the magnificent luxury of
L. Verus, in Capitolinus' estimation, was the possession of
a Crystal bowl, called *Volucer*, after a favourite race-horse,

which was too large for any man to drink off, "humanæ potionis modum supergressum." But Mohammed Ben Mansur boldly adduces Teifashi telling of a merchant in Mauritania who possessed a basin made of two pieces of Crystal so large that four men could sit in it at once; and with somewhat more probability, that in the treasury at Ghasna (Ghisneh), when captured, A.D. 1159, were found four Crystal vases, each of which would hold two skins of water.

That these vases were partly exported from India ready manufactured, partly carved into more graceful forms at Alexandria (the sole channel through which Indian productions flowed into Europe), appears from the passage in Martial (xii. 74) mentioning Crystal vessels as brought to Ostia by the corn-fleet of the Nile :—

> " Dum tibi Niliacus portat Crystalla cataplus."

That such "gemmæ potoriæ" were wrought at Alexandria for the Roman market Martial again intimates (xi. 11) :—

> "Tolle puer calices tepidique toreumata Nili,
> Et mihi secura pocula trade manu
> Trita patrum labris, et tonso pura ministro ;
> Antiquus mensæ restituatur honos."

He will have nothing to do with these costly fragile importations of the day, but calls back the old-fashioned ornaments of the table, the plain silver cups worn by family

---

* In the famous dinner given by L. Verus (which cost *sexagies* or 60,000*l.*, though the party did not exceed twelve), at each remove not merely all the plate used was presented to the guests, but also the servants who had waited upon them. The idea was evidently taken from the wedding of Caranus. After each health the feasters kept for themselves the murrhine, and *Alexandrine* crystal cups they drunk out of. (Capitolin. v.)

use. And that he is opposing the old silver plate to the new-fangled gem-vases from Egypt is decided by his conclusion—

"Te potare decet gemma qui Mentora frangis
　　In scaphium mœchæ, Sardanapale, tuæ."

Declaring that a drinking-cup of precious stone was only befitting the debauchee that could melt down old chased plate by the renowned Mentor to make a " chamber-service," in modern parlance, for a present to his mistress.

There is a great similarity in the elaborate designs of arabesque foliage covering these bowls with those seen on the cameo-vases in glass of known Alexandrian fabrique, such as the elegant amphora of the Museo Borbonico, surrounded with a network of vine-branches. Indeed, Achilles Tatius (ii. 3) has " a crater (deep vase), the entire work carved out of one fossil Crystal (in contradistinction to glass), which vine-branches encompass like a garland."

The Arabs kept up the art they found flourishing there of working in Crystal long after their conquest of Egypt. One of the most valued ornaments of the treasury of S. Denys was : " Un vase de Cristal de roche fait en façon de broc (pitcher) avec son anse, le tout d'un piéce ; le couvercle d'or, attaché à un chaisne d'or. Ce vase est orné de feuillages et d'oiseaux perchez sur des branches sous lesquels on voit force *lettres Arabesques,* le tout en relief. Il est fort estimé et admiré tant pour son antiquité (car il a servi au *Temple de Salomon*) que pour l'artifice avec lequel il est taillé. Il vient de l'empereur Charles le Chauve." ('T. de S. Denys,' p. 120.)

Under the Lower Empire, Crystal seems to have been much in use for making solid finger-rings, carved out of one single piece, the face engraved with some intaglio serving for a signet. All those known to me have the shank moulded into a twisted cable; one example bore for device the Christian monogram, which indicates the date of the fashion.* It would seem that these rings superseded and answered the same purpose as the balls of Crystal carried at an earlier period by ladies in their hands for the sake of their refreshing coolness during the summer heats: a fashion kept up by the Japanese to the present day. Propertius (ii. 24 and iv. 3) makes the deserted fair one complain—

" Et modo pavonis caudæ flabella superbæ,
   Et manibus dura frigus habere pila.

   .   .   .   .

   Nam mihi quo Pœnis si purpura fulgeat ostris,
   Crystallusque meas ornet aquosa manus ? "

" Now courts the breeze with plumes of peacocks fann'd,
   Now holds the flinty ball to cool her hand.

   .   .   .   .

   Oh! what avails the purple's Tyrian glare,
   Or that my hands the limpid crystal bear ? "

Such balls, frequently found amongst ancient remains, even of the Ninevites (Dr. Dee's showstone is one of them), were also used as burning lenses. That they were so employed by surgeons is expressly stated by Pliny :—" I find it asserted by physicians that, when any part of the body requires to be cauterized, it cannot be better done than by means of a Crystal ball held up against the solar rays." Long before, Aristophanes (Nubes, 758) jocularly

* The idea of such rings was unmistakeably borrowed from the hemispherical seals in Calcedony of the Sassanians, that national form of the signet: in many of which the large diameter of the perforation and the carving of the sides, render them but little different in appearance from these ornaments of their Roman and Byzantine rivals.

introduces the ϋαλος in its capacity of a burning glass as
the brilliant invention of Strepsiades (under the inspira-
tion of Socrates) for the quashing of a writ :—

> " You have noted
> A pretty toy, a trinket, in the shops,
> Which, being rightly held, produces fire
> From things combustible?
> *Soc.*   A burning-glass,
> Vulgarly called?
> *Str.*   You 're right, 'tis so.
> *Soc.*   Proceed.
> *Str.* Put now the case : your scoundrel bailiff comes,
> Shows me his writ. I, standing thus, d'ye mark me,
> In the sun's stream, measuring my distance, guide
> My focus to a point upon his writ,
> And off it goes in fumo ! "

Orpheus (170) recommends such an instrument as the
most proper for kindliug the sacrificial fire, and ensuring
the favour of the gods, the flame thus kindled being called
the fire of Vesta.

The largest mass of Crystal ever seen by the Romans
was that dedicated in the Capitol by Livia, which weighed
fifty pounds. Juba mentions a piece found in the Topaz
island of the Red Sea, a cubit in length;* and Bocchus
describes specimens of wonderful magnitude found in
Lusitania in the Ammeæ mountains, in shafts sunk down
to the level of the water. But the Indian sort was preferred to every other, although that found in the Alps
was also esteemed. The latter was extracted from rocks so
difficult of access that the miners had to be lowered down
to them by ropes, being guided by certain indications
known to the experienced as to the veins that would
contain the Crystal. Pliny confirms the assertion of

---

* According to the same authority, the " Island of the Dead," off the
Arabian coast, also produced it.

Sudines, that it was only produced in rocks that faced the south, never in a watery soil, even where the climate is so cold that the rivers freeze to the very bottom. The universal opinion as to its origin, that it was ice, hardened by intense frost (gelu vehementiore concreto), gave to *Crystal* its name, the Greek word for ice. Rain, says Pliny, mixed with a small proportion of snow, is necessary to its formation ; and, as a necessary consequence, the substance is unable to bear heat, and can only be used to contain cold liquids.

Pliny (xxxvi. 66) quotes a statement that in India glass was made out of broken Crystal, and hence the incomparable superiority of the Indian glass to any other. Can some faint report of the excellence of the Chinese porcelain (in which pounded felspar is the essential ingredient) have reached the Greek traders and travellers, his main authorities in this section of his great work ?* for it is certain that glass was unknown in India until imported by the Portuguese ; whilst, on the other hand, Chinese porcelain, when first brought to Europe, was considered a species of glass. It is, however, a fact that Crystal, as being the purest form of Silica, supplies the best possible material for glass-making ; on which account it was largely used in the manufactories at Murano in the flourishing times of the Venetian glass-trade, and now constitutes the chief ingredient of Strass, the basis of artificial gems. Theophrastus (49) refers to the same belief thus : " For if glass be made, as it is said, out of the Hyalites, this article also is produced by condensation." Now the same stone apparently is enumerated by him (30) in the list of signet-gems, where, however, it is distinguished from the Crystal,

---

* A recent traveller, who had penetrated very far into the interior of China, beheld to his great amusement a common " black bottle " (Lord Cardigan's *bête-noire*), holding the post of honour granted to its rarity, and figuring as the apex to a pyramid of the most precious porcelain of the national fabrique.

and termed the Hyaloeides, which may mean the Beryl, a gem not mentioned by him under that name. For, strangely enough, Garcias ab Horto asserts " that no Crystal at all is found in India, but, on the contrary, the Beryl in large fragments, out of which the natives make both *glass* and vessels of price." The first part of this statement is altogether erroneous, for Crystal is still brought from India, and in masses of enormous size.

Glass had been carried to such perfection when Pliny wrote as to imitate the Crystal with wonderful exactness, yet, what surprised him, vases of the latter substance had risen in value, instead of declining through this competition. This colourless, transparent glass, approaching as nearly as possible to the true Crystal, was then the most admired. Such perhaps was the material of the two small vases with handles (pterotæ), a discovery of Nero's times, which sold for the enormous price of 6000 sesterces (60*l.*), for Pliny quotes them when speaking of this stone.

The Indians (xxxvii. 20) had discovered the art of forging all the coloured gems, but more especially the Beryl, by staining Crystal. Treatises were extant, says Pliny (75), directing how to stain Crystal so as to pass for the Emerald and other transparent precious stones; but he declines to point them out, on the ground that even luxury ought to be protected against fraud; adding that no other mode of cheating in the world was so lucrative. Dutens, however ('Pierres Précieuses,' p. 67), is less scrupulous; stating that a Crystal made red-hot and plunged repeatedly into the tincture of cochineal becomes a Ruby; if into tincture of red santal, it takes a deep red tint; into tincture of saffron, a yellow like the Topaz; into that of turnsole, it assumes the colour of the Sapphire; into juice of nerprun, it takes a deep viole like the Amethyst; and into a mixture of tincture of turnsole and saffron, it becomes an imitation of the Emerald. Crystal may also be coloured by the dry method, by placing the pieces upon orpiment and arsenic mixed together in a crucible, and so exposing them to heat. The cold method consists in steeping the Crystal in oil of turpentine saturated with verdigris, or spirits of wine holding dragon's blood or other coloured resins in solution, the depth of tint produced being proportioned to the time of steeping. Such false Rubies are still known to the French lapidaries according to Barbot by the name of *Rubaces.* In London they figure as " Rubies of Ancona." But the substance of the Crystal is not really coloured ; it is merely the dye, which, insinuating itself into the innumerable minute fissures produced by the sudden cooling, that apparently stains the entire mass ; and the great art in the process is so to regulate the heat that these fissures be not too apparent upon the surface of the imitative gem.

It is, however, not unlikely that the ancients employed also the more simple method now so much in use, and which produces most of the Carbuncles seen in the London shops, viz., after cutting the Crystal into shape, then to paint the reverse with the required colour, and so to set it with a backing. The fact that anciently gems were for the most part set in such a manner would greatly favour the execution of this fraud, to baffle which Pliny expressly notices that the better kinds of Jaspis and Chrysolithus were mounted à jour, " funda perspicua." Although the Roman jewellers made up the false Sardonyx of three layers by cementing together as many slices of different coloured stones, yet they do not seem to have been acquainted with *Doublets,* that favourite device with the modern trade, in which a thin slice of the true gem, improved by cementing a paste of the proper colour underneath, assumes the appearance of a first-class stone in its

(G) I

kind; or where, the operation being reversed, the paste surface is backed by a facetted Crystal which serves to supply the required lustre. The ancient frauds in the matter of coloured gems went no farther than the above-named staining of the Crystal, and the substitution of pastes for the true stones. To detect the latter trick, Pliny lays down (76) many rules, some fanciful enough, but amongst them one infallible test, that a splinter of Obsidian will scratch a paste but not a real gem.

Fittingly to conclude the subject of False Gems, which naturally falls under the head of Crystal, so much employed, either in its natural form, or chemically metamorphosed, in their composition, we may insert some of the curious observations of Camillo, made before 1502, as to the various forgeries practised by the Italian jewellers of his own times. Many of these are extremely ingenious, and doubtless the recipes had been preserved in the traditions of the craft from the date of the Empire. Besides making pastes of *smalto* (enamel) exact counterfeits of the true gems, they converted the commoner stones into others of the most precious kinds, by such illusive processes as that of cutting a Garnet very thin and backing it with Crystal to pass for a Ruby; an Amethyst hollowed out and filled with a coloured tincture imitated the Balais, which was also counterfeited by a thin table of the former laid upon a ruby-foil. Diamonds they forged by cutting the Beryl or the White Sapphire to the proper shape (a pyramid), and backing them with the customary tincture. To explain the latter term, it must be borne in mind that formerly the Diamond was always coloured black upon the *culasse*, with a certain composition, on the proper nature of which Cellini treats at great length in his ' Orefeceria,' as being of vital importance to the effect of the gem. In order to escape the test of the file, which no paste can endure, these early Italian forgers imitated above all the rest the Emerald and the Peridot, for, these gems being but little superior in hardness to paste, such a means of detection could not be safely applied to them.

Although it has been remarked above that no antique intagli in this stone are known to exist, yet there is an epigram by Diodorus (Anth. ix. 776), entitled, " Upon an engraved Crystal."

> " The art and colouring Zeuxis well might claim,
> Yet Satyreius is my author's name,
> Who in the crystal drew Arsinoë's form,
> A faithful image with life's beauty warm :
> An offering to his queen : though small in size,
> No larger work with me in merit vies."

This was an image of Arsinoë " drawn " by Satyreius, to which the poet ascribes all the " colouring and the elegance " of Zeuxis, strange expressions for a work in a gem. But the term δαίδαλον* applied to it shows that he is speaking of a bust in full relief, work of which examples in Rock-crystal are not wanting, the most famous being the Odescalchi Venus, nearly six inches high, and executed in the best antique manner ; and, what bears more directly upon the point, the fine Memnonian head in high relief of a young Ptolemy lately acquired by the British Museum.† On the other hand, it was the material which the Renaissance masters preferred above all others

---

* Δαίδαλον properly signifies a work in relief: thus the brooch of Ulysses (Od. xix. 227) is described as παροιθε δε δαιδαλον ηεν, which was a hound seizing a fawn : evidently a repoussé chasing in gold plate, of which so many Etruscan ornaments supply examples. Thus were termed the wooden figures made by the Platæans at the festival of the Dædala, and Pausanias thinks that the old Dædalus got his name from his profession of statuary.

† Plutarch enumerates, amongst the other extravagancies of Cleopatra, her sending billet-doux, " engraved on Crystal and Onyx," to Antony, when seated upon his tribunal in the forum of Alexandria.

for their most important works. Their intagli were
not, like those of the ancients, confined to the small
dimensions of signet-stones, but were designed to be worn
as medallions, to be set in plate, or to form the panels of
reliquaries or caskets (being laid upon a coloured ground),
for all which uses extent of surface and transparency were
the greatest recommendations. The fame of Valerio, il
Vicentino, chiefly rests upon his large plaques of Crystal
engraved with elaborate compositions done in his own
peculiar and forcible style, of which numerous specimens
(due to his incredible industry through the course of a
long life), are everywhere to be seen in collections. The
most important of these was the coffer adorned with the
History of the Passion, a commission from Clément VII.
at the price of 2000 gold scudi, designed for a present to
François I., on their interview at Marseilles, where they
arranged the marriage of the Dauphin with Catarina dei
Medici. Vasari especially commends the Crystals of
Valerio's rival, Gio. del Castel Bolognese, particularizing
as his best the Tityus, after the design of M. Angelo, and
the Ganymede engraved for the Card. Ippolito dei Medici.
Another famous work of his was the Attack on the Castle
of Bastia, done for Duke Alfonso of Ferrara. A Lion-hunt
by this artist, now in the Devonshire Collection, fully
justifies the encomium of Vasari. These plaques are
usually of considerable size, one of Valerio's in the Uzielli
Collection measuring 3¼ × 2¼ inches. Mariette says that
the Tityus and the Fall of Phaethon were executed after
drawings furnished by M. Angelo, and suspects that Perin
del Vago supplied those for his scenes from the Life
of Christ adorning the bases of the Cross, and two
candlesticks presented to St. Peter's altar by the Card.
Alessandro Farnese, the engraver's latest and chief patron.

Smaller intagli of this date in Crystal are sometimes

found in jewelry set with the engraved side downwards
upon a gold or an azure foil. The effect thus produced is
very singular, the figures appearing as though cut in relief
upon a transparent gem, a Topaz, or a Sapphire, and the
illusion so perfect as only to be detected by the touch.
A veiled bust of the Madonna thus treated and set in a ring
which lately came under my notice was a remarkable example
of this curious effect, the head appearing in high relief,
though upon a really plane surface. This whimsical application of engraved Crystals seems to remount to the last
ages of the Lower Empire; certain it is that we find it the
general rule in Carlovingian art. Of this the most remarkable examples are the flat circular plaque, six inches in
diameter, which served from time immemorial for the
morse to the vestments of the Abbot of Vézor sur Meuse.
It is covered with numerous groups, representing the story
of Susanna and the Elders, engraved in a peculiar and
slight intaglio, and explained by neatly-cut legends underneath each. The reversing of the letters in these and in
the central medallion, "*Lotharius Rex me fieri fecit,*" demon-
strates that the subject was intended to be looked at
through the *back* of the crystal plaque. This precious
monument of the expiring Glyptic art is now mounted in a
silver-gilt frame in the later Gothic style, but which ever
passed, like innumerable other ecclesiastical jewels, for the
genuine work of St. Eloi. The other piece of the same
period is unusually large, being 7 × 3½ inches in length
and breadth, and very deep from the extreme convexity of
its back. The subject is the Crucifixion : at the foot of the
Cross writhes the bruised Serpent ; Mary and John stand on
each side in flowing Anglo-Saxon drapery. The *technique* of
the engraving indicates clearly that it came from the same
school, perhaps from the same hand, as the preceding work.
The convexity of the reverse, which is highly polished,

enormously magnifies the figures on the flat surface beneath, when seen through it. That this was the object of the Frankish goldsmith in thus employing such intagli in ornamentation, appears from the fact that the famous Beryl of Evodus was similarly reversed in the " Escrin de Charlemagne ;" and from the optical illusion thus designedly created, the head of Julia is described by those that first mention it as being in relief. Both the plaques above quoted have fortunately been secured for the British Museum within the last few years.

From the very commencement of the Revival, Crystal was employed once more for its ancient and proper destination, the formation of vases in the most elegant and varied forms ; as early as the close of the fifteenth century, Camillo particularizes this as the sole use of the material. This branch of industry, so eminently artistic in its character, flourished most vigorously for the two following centuries, as the sight of the older European cabinets fully attests ; and its productions, when remarkable for magnitude or skill, commanded then (and even now) the highest prices. The works in Crystal belonging to the Crown of France are valued in the inventory drawn up in 1792 at one million of francs (40,000*l.*), comprising vases of numerous different shapes, fifteen candlesticks, statuettes, a Death's Head, &c., and a ball ·165 m. in diameter (6½ inches) valued at 10,000 francs ; a galley, ·335 m. (12 inches) long, at 24,000 francs ; and a square coffer formed of six plaques, at 4000 francs.

Such objects will find their value augment with the progress of time in the same proportion as their rarity ; for in our days this elegant art is entirely extinct. It has been completely driven out of the field by the invention of Strass, of Flint-glass, and of artificial Crystal ; substances superior to the natural in homogeneousness, purity, and lustre, and only yielding to it in point of hardness. Although the refractive power of the Crystal is but a third of that of the Diamond (being as 10·5 to 30), yet the Cinque-cento jewellers had found a way so to cut it into a pyramid, that, when mounted in the favourite " tower-rings " of that period, it presented the iridescence and much of the lustre of the gem it was designed to counterfeit. De Laet mentions the having seen " Cornish Diamonds " cut into " Tables " that could not be known from the real gem.

Crystals are sometimes found with a cavity in their sub-stance containing a few drops of water, which moves about as the stone is turned. This is Pliny's Enhydros or Enhy-gros. "always perfectly spherical, colourless, and polished, but, when moved, something fluctuates about within it like the liquid in an egg." This was regarded by the ancients as a most wonderful miracle of nature, and an irresistible proof of the correctness of the theory deducing its forma-tion from hardened or solidified ice. Claudian has left several elegant epigrams all turning upon this idea :—

" Pass not the shapeless lump of Crystal by,
Nor view the icy mass with careless eye ;
All royal pomp its value far exceeds,
And all the Pearls the Red Sea's bosom breeds.
This rough and unform'd stone, without a grace,
Midst rarest treasures holds the chiefest place."

" With th' Alpine ice, frost-harden'd into stone,
First braved the sun, and as a jewel shone,
Not all its substance could the gem assume—
Some tell-tale drops still linger'd in its womb.
Hence with augmented fame its wonders grow,
nd charms the soul the stone's mysterious flow :
Whilst stored within it, from Creation's birth,
The treasured waters add a doubled worth."

" Mark where extended a translucent vein
Of brighter Crystal tracks the glistening plain :

No Boreas fierce, no nipping winter, knows
The hidden spring, but ever ebbs and flows :
No frosts congeal it, and no Dog-star dries ;
E'en all-consuming Time its youth defies."

" A stream unfetter'd pent in Crystal round,
A truant fount by harden'd waters bound :
Mark how the gem with native sources foams!
How the live spring in refluent eddies roams !
How the live rainbow paints the opposing ray,
As with the imprison'd winter fights the day !
Strange nymph ! above all Naiads' fame supreme—
Gem, yet no gem—a stone, yet flowing stream ! "

" Erst while the boy, pleased with its polish clear,
With gentle finger twirl'd the icy sphere,
He mark'd the drops pent in its stony hold,
Spared by the rigour of the wintry cold.
With thirsty lips the unmoisten'd ball he tries,
And the loved draught with fruitless kisses plies."

"Streams which a stream in kindred prison chain,
Which water were, and water still remain,
What art hath bound ye, by what wondrous force
Hath ice to stone congeal'd the limpid source?
What heat the captive saves from winter hoar,
Or what warm zephyr thaws the frozen core?
Say in what hid recess of inmost earth,
Prison of floating tides, thou hadst thy birth ?
What power thy substance fix'd by icy spell,
Then loosed the prisoner in his lucid cell ? "

This singular phenomenon equally excited the admiration of the mediæval philosopher, Marbodus devoting a separate section to a description of the mystery which he owns himself unable to explain. It is the modern opinion, however, that the liquid, instead of being imprisoned in its "lucid dungeon" at the time of the Crystal's formation, and being the residuum of its subject-matter, rather has subsequently infiltrated into the cavity through the pores of the stone. It is said that the miners in California often come upon huge quartz-nodules filled with water, and are poisoned occasionally by drinking it, from the quantity of silica it holds in solution.

Such Enhydrous Crystals are not unfrequently to be met with in jewels of the Cinque-cento. They are, indeed, somewhat dangerous ornaments, being apt to explode with great violence on the sudden application of heat, an instance of which is on record where a person had his palate severely lacerated by inadvertently placing one in his mouth and its bursting there; and Caire mentions a similar explosion happening within his knowledge on a jeweller's attempting to solder a ring set with such a "pregnant gem." *

The ancients do not appear to have attached much medi-. cinal virtue to the Crystal—Orpheus, the prime source of all such notions, only recommending it as a remedy for diseases of the kidneys, externally applied, and as a burning lens for sacrificial purposes; but Marbodus, on the authority of Evax, prescribes the powder of it to be taken in *mulsum* (wine and honey) by women suckling, as a sure means for increasing their supply of milk.

In addition to all these wonders, the Crystal played an important part in magic. A sphere, some three inches in diameter, apparently one of that ancient sort already described, was the far-famed "show-stone" of Dr. Dee, wherein that "egregious wizard" conjured up for those who consulted him a picture of their coming fates, after the manner yet practised by the Caireen magicians, who, however, substitute for the sphere a pool of ink held in the

* These curious stones are now found at Bragonza near Vicenza, whence their popular name "Goccia d' aqua di Vicenza."

palm of a little boy's hand. This magic mirror may be
seen preserved in the class of Crystals (quartz) in the
British Museum. Dr. Dee had very ancient authority for
such a mode of foreshowing the future, for Marbodus thus
transcribes the precepts of some antique *magus*, though
giving the *Diadochus* (simply mentioned by Pliny as resembling the Beryl) for the Crystal :—

> " If e'er thou seek, where deep the rivers flow,
> To force the water-sprites the fates to show,
> Take the *Diadochus* upon thine hand ;
> No gem more potent o'er the fiendish band.
> Within its orb to thine affrighted eyes
> Shall myriad shapes of summon'd demons rise.
> But mark ! if brought in contact with a corse,
> Forthwith the gem shall lose its native force.
> Like to the Beryl shows the wondrous stone,
> That dreads the touch of one by death o'erthrown."

This mode of prying into the book of Fate had been
established from time immemorial with the ancients. Spar-
tean records that Didius Julianus in his despair, " besides using unhallowed spells, resorted to what is called
divination by the mirror, wherein a boy blindfolded looks,
after a charm has been uttered over his head :" adding
that the boy descried in this manner the coming of Severus,
and the death of Julianus.

CYANUS: Κυανὸς: *Lazulite.*

I⊤ has been asserted positively by some modern writers
(Millin, &c.) adopting the conjecture of De Boot, that
Pliny's Cyanos was our Sapphire; an opinion, however, by
no means borne out by his description of its natural characters. "The Cyanos shall be described separately, a
favour granted to the blue colour lately mentioned (*i. e.*
when speaking of the Blue Jasper). The best sort is the
Scythian, then the Cyprian, and last of all the Egyptian.
It is very largely imitated by *staining crystal*, and a certain
king of Egypt has the credit of having first discovered
how to tinge Crystal of this colour. This stone also is
divided into the male and female kind. Sometimes there
is gold-dust seen within it, but differently from that in the
Sapphirus: for in the latter the gold shines in points or
specks amidst the azure."

Pliny appears to have written the above, not from a
knowledge of any gem going by that name at the time, but
merely as a translation, somewhat inaccurate, of the words
of Theophrastus (55):—" And as there is a Red Ochre both
natural and artificial, so is there a Cyanos also, both produced naturally and made by art, like that manufactured
in Egypt. Of this Cyanos there are three kinds: the
Egyptian, the Cyprian, and a third the Scythian. The
Egyptian is the best for thick-bodied paints, but the
Scythian for those of a diluted kind. The Egyptian is
produced artificially, and the writers of the history of their

monarchy record this also, which of the kings it was who made a *fused Cyanos* in imitation of the natural stone; and that this mineral used to be sent as a present from other countries. From Phœnicia, however, it was brought as an appointed tribute, a fixed quantity of Cyanos, so much in the native state, and so much ready calcined. The persons who grind up paints say that the Cyanos produces from itself four different shades of colour: the first made from the thinnest pieces being the palest, the second from the thickest giving the deepest tint."

From all this it is evident that Theophrastus is describing some mineral used merely as a paint, and not as a precious stone,* and that his Cyanos was nothing else than an inferior Lapis-lazuli, like what we see in antique Egyptian jewelry, or rather a Lazulite prepared by calcination for the painter's use, that is converted into *ultramarine*. Or it may have been Sulphate of Copper in bright-blue crystals (Cyanose): which is nearly transparent in its native state, and of considerable hardness: or again something of the nature of the Lapis Armenius (Berg-blau) noticed by Tavernier and Chardin as employed in Persia in their own times as a cheap substitute for ultramarine. The gold-dust, however, noticed by Pliny as occurring upon its surface, though but occasionally (of which Theophrastus says nothing), would again lead to the conclusion that *his* was a variety of the Lapis-lazuli, pieces of which sometimes are found entirely filled with the specks of Pyrites. Pliny, understanding Theophrastus to be describing a precious stone, has mistaken his meaning as to the discovery of the Egyptian king, and has rendered his "fused Cyanos" by "stained Rock-crystal," that being the

* The *Sapphirus*, or Lapis-lazuli, is described in the first division of his book, containing Stones and Gems; but the *Cyanos* comes in the second, treating of Earths used in the arts.

common method (learnt from the Indians) of forging the transparent precious stones in the age of Pliny. This mention of Crystal, and the idea consequently involved of the imitation of a blue, transparent, lustrous body, though merely based upon a mistranslation, was the sole foundation for · the belief that the Cyanos could be the modern Sapphire, which the whole tenor of the statement of Theophrastus proves conclusively it was *not*.

The artificial Cyanos is the blue enamel, as our Museums show, used by the old Egyptians for glazing all their works in terra-cotta, in imitation of the more sumptuous articles in the then so precious Lapis-lazuli. Davy found that the Egyptian azure, employed in fresco-painting, could be exactly and cheaply produced by *fusing* together during two hours 15 parts carbonate of soda, 20 powdered flint, and 3 of copper-filings, yielding thus the true artificial Cyanos of the Greek mineralogist. The chemist will perceive from its ingredients that this composition could be applied as an enamel, as well as employed in powder as a paint; and that *flint* replaces the *crystal* of the old Egyptian recipe.

There can be no doubt that the Cyanos of the earlier Greeks was the mineral known to Pliny as *Armenium :* both names being mere epithets, the first denoting its colour (blue), the second its native country. " Armenia supplies the paint called by its own name. It is a stone coloured like the Chrysocolla, and the best quality, that which approximates nearest to that mineral, yet mixing its colour with a sky-blue. It is usually priced at 12 denarii per pound. That found in Spain is in the form of sand, but is prepared for use in the same manner. The cheapness of the latter brings it down to 6 denarii. It differs from Azure (Cæruleum) by reason of a slight tinge of whiteness, which renders this colour more tender. Its

sole use in medicine is for promoting the growth of hair,
especially of the eyelashes" (xxxv. 28).

But by a curious oversight Pliny has (xxxiii. 57) applied
the exact words of the remainder of the same chapter of
Theophrastus, describing the Cyanos, to his *paint*, Cæruleum, for, after stating that it is found in silver-mines in
the form of sand, he adds: "*Of old time* there were three
kinds of it,—the Egyptian the most esteemed; the Scythian;
this is easily dissolved, and in grinding up divides itself
into four shades—a lighter, a darker, a thicker and a
thinner—to the latter the Cyprian is even now preferred.
To these are newly added the Spanish, and that from Pute-
oli, for they have begun to manufacture it in those places.
It is all, however, stained and boiled with a peculiar plant
until it imbibes the colour: afterwards it is treated like
the Chrysocolla. The test of the purity of Cæruleum is for
it to blaze if thrown upon hot coals; the false kind is made
by mixing the water in which dried violets have been
boiled with Eretrian chalk. The effect of Cæruleum in
medicine is to cleanse ulcers, on which account it enters
into the composition of certain ointments." All which
notices go to prove that the base of the several blues of the
Romans was Sulphate of Copper under various forms. And
to sum up: both κυανὸς and Cæruleum are nothing more
than literal translations of the Persian *Azul,* "blue:" a mere
epithet of colour.

EUMITHRES: MITHRAX: *Amazon-Stone: Labrador Spar ?*

A VERY compact felspar, of an emerald-green colour, but opaque and with a nacrous reflexion, excessively brittle but susceptible of a fine polish. The name is derived from the circumstance of its first discovery by the Spaniards amongst the ornaments of the Indians dwelling upon the river Amazon. Like that equally obdurate material the Jade, this stone was a favourite vehicle for ancient Aztec art. Grotesque figures cut in it of idols and of beasts, wonderfully rude in design, but very skilfully carved and polished, are not unfrequent in collections, having been exhumed from the tombs of the old Mexicans. All these carved objects are perforated for a string, and thereby show that they were intended to be worn round the neck like the similar pendants of sacred animals with the ancient Egyptians.

Ximenes, a contemporary of Cortez, mentions (iv. 18) this stone and the value set upon it by the Mexicans: "They entitle *Hoitzitzettetl,* which signifies 'the precious stone,' a certain gem which some term *Oio di gatto* (Cat's-eye), but by us Spaniards it is named, I believe, the *Tournesole.** It

---

* That is, the heliotrope, for which its colour caused it to be mistaken. The comparison to the *cat's-eye* proves he is speaking of a felspar, otherwise I should be inclined to make the Amazon-stone his *Thayotic* "cut by the Mexicans into figures of men, idols, and round long columns," and which the Spaniards called *pietra di hijada.*

is found in great plenty in the province Toltotepec, but the stones there are small in size, and on account of their extreme abundance are but little valued."

It has been of late years ascertained that this curious mineral was known to the ancient lapidaries both Oriental and Roman, and that, too, at the very dawn of the Glyptic art. · The skilfully-engraved Assyrian cylinder assigned to Sennacherib (in the British Museum) is described as cut in Amazon-stone; and certain camei, particularly a veiled head of M. Aurelius (Praun), in a stone the character of which could not be mistaken, have come under my own observation. Lastly, the question is set at rest by the fact recorded by Corsi, that in some excavations carried on in the ruins of the villa of M. Vopiscus at Tivoli, by Lord Northampton (1826), were discovered fragments of the pedestal either of a statue or a column sculptured in the true Amazon-stone. These fragments were covered with *hiero-glyphics*; a fact that, coupled with the necessary large dimensions of the mass, throws an unexpected light upon the nature of the monster *Egyptian* Smaragdi quoted by Theophrastus, and removes their existence from the domains of fable. Corsi, indeed, goes on to identify this stone with Pliny's *Chalcedonius*, on the score of the nacrous reflex that characterized the latter, but the minuteness and fusibility of the crystals of the Chalcedonians are insuperable objections to this theory.

The above-quoted cylinder recalls to one's memory two gems both described by Pliny as productions of Persia, and both seeming, from his brief notice of their properties, to present in different points certain analogies to those of the Amazon-stone. His first is the *Tanos:* "in colour a disagreeable green, and which is foisted in amongst the Smaragdi;" and certainly the cameo above alluded to had so much the appearance, that it was described in the catalogue by Urlichs, Professor of Mineralogy at Bonn, as actually executed in Emerald. Pliny's second is the *Eumithres,* "or gem of Belus, of the colour of a leek-leaf, and a favourite in their superstitions,"—a character that would give its bearer the strongest claims to the honour of serving for the signet of the King of kings.

Before dismissing the gems belonging to the Felspar class, a word must be given to the most beautiful of them all, the Labrador-spar. Pliny briefly defines a certain stone found in Persia and the mountains upon the Red Sea, named the *Eumithres,* "favourite of Mithras," as many-coloured and reflecting divers tints when held against the sun. This would serve for a loose description of the Labrador, of a silvery-gray when laid horizontal, but when inclined to the light, flashing back an exquisite iridescence in which azure predominates, covering all its surface.

This stone is found on the coast of Labrador, and the finest specimens lie under water, where they form the flooring of the shallows, and betray themselves by their lustrous reflexion to those passing over them in boats.

GAGATES: Γαγάτης: *Jet.*

THIS stone (or rather fossil wood) derived its ancient name
from the river Gages in Lycia, where it was first discovered;
it was also said to be picked up on the coast of Leucolla
over an extent of twelve furlongs, where it had been cast
up by the waves. The Romans chiefly valued it for its
supposed medicinal qualities, some of them very wonderful,
for the fumes of it when burning would discover any one
subject to epilepsy by immediately inducing a fit: and the
water in which it was steeped proved by its undisguisable
effect an infallible ordeal for female chastity. Its fumes
also drove away all reptiles, and were good against the
strangulation of the womb. Its flames, though quickened
by water, were quenched by the application of oil. Anything painted with it upon pottery* could not be obliterated.
Mixed with wine it was good for the toothache; and with
beeswax, a sovereign ointment for tumours. But strangest
virtue of all was that for which the Magi employed it in
the mode of divination called *Axinomantia*, for it remained
incombustible in the fire, if the desire of the consulting
party was destined to be accomplished. The same virtues

---

* A remark proving that indurated Bitumen was then confounded
with the true Jet, for the former was the base of the black pigment so
extensively employed by the Greek and Etruscan potters in the decoration of their painted vases. In our times likewise, the so-called
Russian Jet is the same mineral, and is from its nature much used for
making black sealing-wax.

are set forth by Orpheus (468), who indeed appears to have been Pliny's authority on this article. Marbodus notices the electricity it acquires by friction, and the fact that the British kind was of equally good quality with the Lycian. This electricity procured for it in the Middle Ages the title of Black Amber; in fact, it often occurs in the same beds of Lignite as the real Amber, and is probably due to the fossilised branches of the same tree that produced the resin, the origin of the latter. This origin is declared by the regular woody structure Jet presents when cut in very thin slices and viewed by transmitted light: it then becomes translucent, and changes to a reddish-brown. When the Roman traders brought back the tale of the natives on the Baltic coast employing Amber for fuel, ignorant of its value, it is allowable for us to modify the statement, and interpret their report as relating to the use of coarse Jet, or Kimmeridge Coal a cognate substance, for such a purpose. The earliest Celtic remains would serve to show that Amber was from the first too much valued as an ornament ever to have been thus wasted.

Jet was, however, turned by the lathe into ornaments by the Britons, perhaps even before the Romans subjugated this island, since large rings worked out of solid pieces, for bracelets and anklets, are often· discovered amongst other British remains. The round disks, cut out from the centre of these rings, the refuse of the turner, often found in heaps together in Dorsetshire, long puzzled antiquaries, who agreed to call them "Kimmeridge Coal-money," and to regard them as a primitive currency! Their true origin has been but lately ascertained.*

The Roman settlers in Germany and Gaul learnt from

* It will be observed that Pliny was unacquainted with the use of Jet as an ornament, merely specifying its value in medicine and in the arts.

their subjects how to convert this elegant material into
articles of decoration stamped with their own more refined
taste ; most interesting examples of which were discovered
in two stone coffins found at some depth beneath the
principal entrance of S. Gereon, at Cologne, during the
repairs going on there in 1846. These are conjectured to
have been a complete set of ornaments belonging to a
priestess of Cybele, and consist of two hair-bodkins, with
heads formed out of pine-cones, almonds (beads), trefoils,
bracelets, rings, a half-crotalon with a Medusa's head upon
it,—in all, 26 articles.

The intagli in Jet however, antique and mediæval,
palmed off in such quantities of late years upon English
antiquaries, are nothing but, impudent recent forgeries,
with no ancient precedents whatsoever.

It is not altogether foreign from this subject to adduce a
fact of which few are aware, that the use of *Coal* for fuel
can be traced back to the early times of Greece. Theophrastus, treating of the Anthrax (16), has : " But those
which are properly named *Coals*, on account of the use
made of them by the luxurious, ignite and burn exactly
like charcoal. They are found in Liguria in the same
region as the Amber ; and also in the territory of Elis, on
the mountain-road to Olympia ; and these are used by the
blacksmiths for fuel."

HELIOTROPIUM: *Heliotrope and Bloodstone.*

THIS stone retains its ancient name, and is a Prase, or semi-transparent, green Calcedony interspersed with small patches of opaque, bright-red Jasper. It is Pliny's variety of the Prasius that "sanguineis punctis obhorret;" "is horrent with blood-red particles." India then, as now, was the only source of this beautiful stone; the best specimens of which present a pleasing contrast of a true Emerald-green nearly transparent, with the most vivid red.

This species must be distinguished from the Bloodstone (*Jaspe Sanguin*) also often, inappropriately, termed Helio-trope, the latter being a green Jasper, perfectly opaque as to both its constituent colours; and besides, a much com-moner material, being found in many parts of Europe. Lastly, we have the rare Tiger-agate of Malwa, only differing from the Heliotrope in being spotted with bright yellow.

The origin of the name, literally "Sun-reflector," is thus given by Pliny: "The Heliotrope, produced in Ethiopia, Africa, Cyprus, is of the colour of the leek-leaf (like the Prasius), and marked with veins of blood. The reason for the name is because, if thrown into a vessel of water, it (especially the Ethiopian kind) transforms the sun's rays falling upon it into the reflexion as it were of blood (san-guineo repercussu). The same stone, out of the water, acts as a mirror for observing and detecting solar eclipses, by showing the moon passing over his disk."

The last notice is very curious, referring as it does to some method used by the ancient astronomers in making solar observations, for it must be borne in mind that the chief authorities upon the properties of gems quoted in Pliny's *Alphabetical List* (in which the Heliotrope stands) were the Magi, the fathers of astronomy. It is, however, difficult to conceive how a tablet of Heliotrope could have been used as a mirror; but, on the other hand, a thin slice of a green transparent stone would admirably protect the eye of an observer looking *through* it. Perhaps here, as in the case of Nero's Emerald (SMARAGDUS), there may lurk a confusion of ideas, between the looking *into* a mirror, and the looking *through* a transparent medium.

Pliny hereupon quotes, as a most glaring example of the impudence of the Magi, their assertion that the wearer of this gem coupled with the plant of the same name was enabled, by means of certain spells known to them, to make himself invisible. To this notion Dante alludes where he sees the damned running about under a hail of fire,—

" Senza sperar pertugio o eliotropio "—

" No hope of hiding-hole or Heliotrope."

Pliny goes on to add that the Hephæstites (Vulcan-stone) found near Corinth might similarly be used for a mirror, though *fiery-red* in colour. The test of its genuineness was if boiling water immediately became cold upon the stone's being thrown into the caldron; or if the same, being placed in the sunshine, forthwith kindled dry fuel. It is impossible to divine what mineral is here intended. Orpheus, however (264), ascribes exactly the same two antagonistic qualities to his Lychnis or Balais-ruby. The name therefore may only be an epithet for the latter, or the Garnet, whose hue the description " rutila " will perfectly suit. In fact,

the locality mentioned as producing it leads to the infer
ence that the Hephæstites was the common ἀνθράκιον of
Theophrastus, found according to him in that place.
Marbodus adds to the other powers of the Heliotrope that
of bringing together the clouds and evoking tempests :—

> " The Heliotrope, or ' gem that turns the Sun,'
> From its strange force the name hath justly won ;
> For set in water opposite his rays,
> As red as blood 'twill turn Sol's golden blaze ;
> And, far diffused the inauspicious light,
> Doth with the strange eclipse the world affright.
> Boils next the vase, urged by its secret power,
> And flings far o'er the brim the sudden shower ;
> And as when day enshrouded is in storms,
> With blackest clouds it heaven's fair face deforms."

This material was very little employed in ancient art
in spite of its beauty ; the irregular distribution of its
colours totally destroying the effect of any work cut upon
it. For, except in the case of the banded Agate, it was a
necessity with the ancient engraver that the field of his
design should be of one uniform shade. A Mercury
seated, the tortoise upon his hand (once Bishop Horsley's) ;
and Sol standing with his whip raised, are the only specimens of really antique work in Heliotrope within my
knowledge, and in both instances we may suspect the substance was recommended to superstition by the analogy of
its name to the nature of the subjects engraved upon it.
The coarser and commoner Bloodstone, though never employed for the productions of high art, occurs not unfrequently amongst the talismans of the later Egyptian
and Gnostic religionists.

The exact converse to the ancient neglect of the latter
stone holds good for the Glyptic art of the Byzantines,
with whom the Bloodstone became the material especially
preferred for their relievi of sacred subjects, particularly

for the half-figures. A mighty recommendation for the substance as the most appropriate vehicle for all such imagery,
was the old tradition that it owed its origin to the stones
lying under the Cross on Calvary, stained by the droppings
of the Saviour's blood on that day, and therefore converted
into imperishable monuments of his sufferings. The same
reason gave it favour with the artists of the Revival, besides the more practical recommendation that in their repre-
sentations of crucifixions and martyrdoms, the sanguine
spots lent themselves readily to their skill for expressing the blood-dripping wounds and lacerations of the
subject. According to Vasari, the first work that brought
the greatest artist of the age, Matteo del Nassaro, into
notice was a Descent from the Cross done in Bloodstone
with so much art that the sanguine spots exactly depicted
the blood trickling from the wounds of the Saviour. It
was purchased by Isabella d'Este, Duchess of Mantua.

The material was, at the same time, greatly admired
when worked into the shape of vases; a moderate-sized
one selling for as much as 200 thalers. De Boot, who correctly defines the distinction between the Heliotrope and the
Bloodstone, speaks of the former as being brought from India
in blocks sufficiently large for sarcophagi, holding a corpse,
to be carved out of them; adding that he had actually
seen such a sarcophagus placed behind the high altar in
St. Donatius' church, Bruges! What Jasper-like mineral
can he possibly have in view?

JASPIS: 'Ιασπις: *Calcedony, Green, and of other Colours.*

GREENNESS and more or less of translucency were the two essential characteristics of the *ancient* Jaspis: "viret et sæpe tralucet Iaspis," is Pliny's definition (xxxvii. 37). To the same effect Epiphanius writes: "The Sixth stone, a Jasper. This in appearance resembles the Emerald, like which it is green, only duller and more opaque, having also its substance green internally like verdigris. There is another sort greener than the sea, deeper in colour and in degree," &c. Hence Orpheus (264) terms it "the Jaspis of vernal hue." The first of the four kinds into which Dioscorides divides his *Jaspis* is "that like the Emerald." The exact stone intended by all these definitions is pointed out with the utmost exactness by Pliny's remark that the variety like the Emerald, but surrounded by an opaque white line, and therefore called Grammatias (or Polygrammos, if by more than one), was worn for amulets in the East. Or, as Epiphanius explains it: "That with four white veins is good to keep off spectres, at least so the magicians tell us." A square amulet on a stone exactly corresponding to Pliny's Grammatias once came in my way: the Emerald-like *paste* of the material was very remarkable. This verdant colour furnished King Polemo * with the

* At first a grammarian, and a protégé of M. Antony's, who made him king of Pontus. I fear, however, the royal poet stole the point of his epigram from a very similar conceit of Plato the Younger's (ix. 747), who flourished B.C. 300.

idea for his pretty epigram (ix. 746), " Upon a Herd of
Cattle engraved in a Jaspis," where they all seem alive,
and are confined within a golden fold to keep them from
straying beyond the grassy mead. And Archias, for another
(ix. 750), " If you saw upon the finger these cows and this
Jaspis, you would imagine that the former were breathing
creatures, the latter a field of grass."

All these details prove that the name in its primary
acceptation signified our *Plasma*, a Calcedony coloured
green by oxide of Nickel, the better quality of which was
distinguished as the " Prasites," " of the colour of verdigris," by Theophrastus, and as the "Prasius" by Pliny
The translucency of the material is apparent from Pliny's
remark that the finer sort were set *à jour* in rings—not
backed, as was the rule for most gems ; also from their
being imitated by glass-pastes, to be detected by " their
throwing out their lustre instead of retaining it within
their substance." A defect of the species, noticed as very
general, was the *sal*, clearly the whitish flaws so often
disfiguring the Plasma.

From all this it appears that the ancient idea of the
Jaspis was exactly the opposite to the modern of the
Jasper ; the latter now designating a class of stones always
opaque, and, in fact, corresponding to the *Achates* of the
Romans.

Subsequently, other translucent colours were admitted
under this denomination ; Dioscorides enumerating, besides
the Emerald-like sort, a second resembling Crystal, a third
Cerulean, and a fourth, the Terebinthizon, " like the Callais."
But Pliny, writing a few years later, widely extends
the list, seemingly making it include several kinds ot
coloured Quartz and Calcedony: one, the most esteemed,
showing a tinge of purple, another of rose-colour, a third
of the Emerald. Again, another resembled the Sard ;

another the violet * (Grecian, our pink cyclamen), definitions applying exactly to the various shades, some bright
and pale, some dull and overcharged with purple, of our
common Amethystine Quartz.

The Persian and Caspian kinds were cerulean in colour,
which gave them their designation of "Aërizusa," explained by Pliny as resembling the hue of the sky on an
autumnal morning : these were also termed Borea (Βορεία)
by the Greeks. That these sorts were our Sapphirine-
Calcedony may, besides their colour, be also inferred from
another circumstance—the finest and bluest specimens of
that stone are chiefly to be met with in the cylinders and
conical seals of the Persians, whose country exclusively
produced it. That brought from the Thermodon was likewise cerulean ; the Phrygian was purple ; the Cappadocian
turning from purple to cerulean (ex purpura cærulea), dull
and lustreless. What species can be intended by these
latter descriptions, unless, as above suggested, it be the
Amethystine or the Rose Quartz, it is impossible to discover.

There were many other varieties of the Jaspis ; the
general fault of them all was the being either cerulean, or
crystal-like (*i. e.* colourless), or resembling the myxa-plum
)a dull yellow?) : all which are peculiarities applicable to
our coloured Quartz-gems. To the same family belonged
the gems called, by way of distinction, " Sphragides," as
the best for making seals—" supremacy over all other
signet-stones having been awarded to them on this account
by public favour." Theophrastus likewise states to the
same effect (23), " that the stones out of which signets are
made for the sake of their beauty alone, are the Sard, the
Jaspis, and the Lapis-lazuli." And, earlier than any,

* I fear Pliny has mistaken here the *ἰὸς, verdigris,* of Theophrastus'
Prasitis for *ἴον,* the *cyclamen.*

Plato has (Phæd. 210), "the stones admired by the Greeks
—Sards, Jaspides, Emeralds"—apparently arranging them
in the order of their value. But here another and an unexpected difficulty presents itself : intagli of the Greek period
are very rarely to be met with on the translucent green,*
or cerulean Calcedonies, constituting properly the Jaspis
of those times. The conclusion therefore cannot be resisted
that the name was also given to those bright yellow stones
(now improperly termed " Amber Sards "), the favourite
material with the Greek engravers, but which went entirely
out of fashion under the Romans. Such a colour the Greeks
termed χλωρὸς, equally with *green ;* for their ξανθὸς meant
orange, being applied to hair which we call *red.* But, in
fact, these "yellow Sards" (to keep to the modern terminology) often have a green tincture, and sometimes
darken into olive, an intermediate shade linking them with
the Plasma.

To the same class must be referred those clear brown
Sards, distinguished in French as " Sardoines," on which
much of the best work in the early styles is to be found.
For this sort the Greeks had an appropriate name, " Capnias " or " Capnites "—the " Smoke-stone," for it has
precisely the appearance of having been thus artificially
tinged. The title " Sphragides " necessarily could only
have been conferred upon a species used for *sphragistic* purposes to the almost entire exclusion of all the rest. The
Greeks did not account them *Sards,* because *red* was the
colour indispensable to the idea of a Sard ; as indeed is
expressed by its Persian original, *sered.* This explanation

* The only exceptions furnished by my experience are the remarkable plasma scarabeus of Dr. Bishop's (Naples), and a ringstone in the
same, curiously mottled by a darker green, engraved with a good figure
of Neptune resting on his trident like the type of Posidonia, inclosed
in an Etruscan border; an interesting gem for many reasons (Mr.
Stowe, Oxford).

serves also to justify the strange epithet " Terebinthizon,"
given by Dioscorides to his fourth subdivision of the Jaspis :
it expresses the limpid, pale, yellow like *turpentine*, that
distinguishes the favourite vehicle of early Grecian art.
Pliny, having in his mind the fraud then so common of
imitating the Sardonyx by cementing together its three
constituent strata by means of Venice turpentine, says that
he fancies the " Terebinthizusa " was called by a name
that did not belong to it (" improprio ut arbitror cogno-
mine "), as being made up out of several stones of the same
family ; but this notion, opposed as it is to the definition
of Dioscorides, cannot apply to an *unstratified* gem. It is
strange that the accurate Virgil should have so far mistaken the primary character of the stone as to describe the
sword of Eneas, Dido's *gage d'amour*, as " stellatus Iaspide
fulva." But the poet had doubtless read χλωρᾷ in some
original he was pilfering, and being no mineralogist, had
understood by the word *yellow* instead of green.*

But what establishes the opinion above advanced, that
the generic name *Iaspis* came at last to include, besides the
Plasma, all colours of the Quartz from purple to yellow
(the Amethystus being our Garnet), is the observation of
Epiphanius, " that there is another sort found in the caves
of Mount Ida, resembling the dye of the purple murex-
blood, but more *transparent*, as approaching to the colour
of wine, and redder than the Amethystus. For," he continues, " the Iaspis is not invariably of one colour, nor of

* His mistake reminds us of Tennyson's use of "Sardonyx" (evidently without the slightest notion of its meaning) when distressed for
a rhyme,

"And the Maid-Mother on a crucifix
　　*　　　*　　　*　　　*
Beneath branch-work of costly *Sardonyx*."

It must not, however, be concealed that Nonnus applies the epithet
ξανθὸς (orange) to the Jaspis in the celebrated necklace of Harmonia.

one quality, but some are softer and paler, and not very
brilliant, nor yet very deficient in lustre.   Others are like
rock-crystal: another kind, somewhat opaque, has *lines* in
the middle; another, called the 'ancient,' is like snow, or
froth of the sea: this last is said by magicians to be feared
by the beasts of the field, and by spectres." And Psellus
so closely follows him in his definition of the Jaspis, that
he must be copying the same original.

Another of Pliny's subdivisions is that " *starred* with red
spots." This is *not* the Heliotrope (which he describes
accurately enough in another place), but a white Calcedony
full of minute sanguine dots, now entitled " St. Stephen's
Stone," and formerly held in high veneration, as thus dyed
ever since its employment in his martyrdom.   Pliny
adds this in at the end of the article amongst a lot of what
evidently are common Calcedonies, without colour, one
containing a cloud,* another tipped with snow, a third like
salt, a fourth smoke-stained, &c.   And as a conclusive proof
that he is talking of the common Calcedony, and of nothing
more precious, he mentions having seen one fifteen inches
high and carved into a statuette of Nero in armour.

His " onychi juncta quæ Jasp-onyx vocatur," is indicated
by the very composition of the name as that extremely rare
Onyx, in which a true opaque red Jasper is superimposed
upon a Plasma, to use the modern terms.   In such a material
was engraved the wonderful Corinthian Helmet, the glory
of the (old) Poniatowsky Cabinet; and Winckelmann particularly notices it as employed for the best works in that
style in Stosch's : a proof of its ancient high estimation.

Pliny speaks of the Jaspis " as still retaining the glory it

---

* There is a specimen of this " Jaspis nubem complexa " in the
Marlborough Cabinet; a clear Calcedony containing a white opaque
spot. The stone is cut into a spheroidal form, and set in a ring ; evidently
from the place it holds being once considered a prodigy of Nature.

had of old, though now surpassed in value by many other gems." And we find that fine specimens continued to be prized under the Empire, notwithstanding the then comparative abundance of the true Emerald. Thus Juvenal excuses the host's watchfulness over the "sharp nails" of his needy guest as he handles the jewelled goblet, on the plea—

"Da veniam : præclara illic laudatur Iaspis."

Similarly Martial, after making his exquisite look over and bargain for all the most expensive wares of the Septa or Grand Bazaar—slave-boys, ivory-carvings, *antique* bronzes, old plate, Crystal and Murrhine vases—ends with his pricing *large* gems of this kind—

"Et pretium magnis fecit Iaspidibus."

It retained its value down to the very latest times of the Roman Empire. Claudian, in picturing Proserpine, has—

"Collectæ tereti nodantur Iaspide vestes,"

and again his imperial bride—

"Viridique angustat Iaspide pectus."

And, last of the Latin poets, Sidonius, in describing the goddess Roma,—

"Ostricolor pepli textus quem fibula torto
Mordax dente vorat; tum quidquid summa refudit
Tegminis hoc *patulo* * concludit gemma *recessu*."

Pliny notices that the finer sort were always set *à jour* : "Præstantiores funda cluduntur ut sint patentes ab utraque

---

* In this epithet Buonarruoti discovers, very plausibly, an allusion to the large Calcedonies carved into masks, and perforated for cords, already noticed under ACHATES.

parte, nec præter margines quidquam auro amplectente."
The extremely globose form given to all antique Plasmas
and Amethysts strikingly corroborates this observation;
showing that they necessarily were intended for setting
without a back.

The conclusion to be derived from all the foregoing
details is undoubtedly this, that the Jaspis of the ancients
was our Calcedony (silica combined with alumina)—in its
primary sense signifying the variety coloured green by
nickel, now called Plasma, but in after times embracing
the blue, the purple, the yellow, the whitey-brown shades
of the same substance; in a word, every colour except the
blood-red, which gave its name and honour to the *Sard*.
This decision is furthermore confirmed by the authority of
Ben Mansur, dividing his *Jasheb* (the original of the Greek
Iaspis) into five kinds: the White clear, the White-yellow,
the Green-black, the Black-transparent, the Dust-coloured:
or in modern parlance, the White Carnelian, the Yellow
and Olive Sard, the Brown Sard (*Sardoine*), and the light
Brown.

The *modern* JASPER, as distinguished from the *modern*
Agate only in its being more opaque and containing more
iron in its composition, was certainly regarded by the ancients as a variety of the Achates. Of this we shall now
proceed to consider the kinds that were chiefly employed
by the engravers of antiquity.

1. First in the class stands the Black Jasper, an extremely fine, close-grained substance, perfectly opaque, and
thus distinguishable from the Black Agate (to which the
ancients were equally partial), which is translucent and
reddish-brown by transmitted light. This stone takes a
high polish, the good preservation of which declares the
uncommon density of its particles. It has been employed
by the Greeks as the material for some of their finest

intagli; for example, the fragment of the head of the Dying
Medusa (Praun), which stands pre-eminent amongst the
innumerable repetitions of that subject. Inferior, or Lower
Empire, work, never occurs in this material. No description of a gem answering to this is to be seen in Pliny's
list; unless it be the one briefly noticed by the name of
*Antipathes,** or perhaps confounded with the *Opsianus.*
The Greeks would probably have considered it an 'Ανθρά-
κιον, the word indifferently signifying a coal dead or alive.
There is nothing but the context to inform us whether in
any particular passage the *Greek* mineralogist is meaning a
flame-coloured and transparent, or a black, opaque substance; whether he is talking of the Ruby or of Jet. And
this want of distinction evidently puzzled Pliny equally
with ourselves: for the Roman *Carbunculus* was a live coal,
and so opposed to *Carbo.*

2. The Red: for which the converse holds good, no *Greek*
intagli existing in this stone, whereas Roman, especially
of the Middle Empire and of the Decline, abound: indeed
the best performances of those times are executed in it.
The most celebrated work in Red Jasper is the Head of
Pallas, taken from the far-famed masterpiece of Phidias,
with the signature of the artist, Aspasius (Vienna). The
earliest portrait in it, known to me, is one of Hadrian; the
subjects, moreover, for which it is principally employed
are in the taste of his age, being chimerae or symbolical
combinations, astrological devices, and Bacchic masks and
scenes. For embodying the last it was recommended by its
colour, vermilion being that wherewith it was customary
to smear the visages of the God of Wine, and of his
attendant satyrs. As this gem came into fashion, and
only begun to be appropriated by the Glyptic Art, after

---

* "Black and not translucent." It was of virtue as an amulet. .

(G)   L

Pliny's date, he nowhere expressly mentions it, though it seems to be intended by some of his descriptions of stones, cursorily mentioned as *red*. Amongst these the *Corallis* has, in my opinion, the best claim to be considered its ancient representative: for "it resembled vermilion in colour, and was produced in India and *Syene.*" The last would explain its frequent occurrence in Egyptian amulets and symbols, such as eyes, fingers, horses' heads, &c. This Syene Red Jasper is pale but uniform in tint, and at first sight would easily be mistaken for unpolished Coral.

This sort may be the "single-coloured Achates" of Pliny, which boiled in oil communicated to it a vermilion colour: indeed Orpheus (609) terms one variety of the Achates μιλτοπάρηος "vermilion-cheeked." It was not the Hæmachates, for *that* is defined by Solinus as "blushing with spots of blood," and therefore was the translucent Blood-agate as distinguished from the opaque, uniformly tinted, Red Jasper. Neither was it Pliny's *stone* Hæmatites (xxxvi. 25), for that was a loadstone, and thereby pointed out as the magnetic iron-ore still retaining the name. With more reason may it be found in the Hæmatitis of Theophrastus (37), reckoned by him amongst the cheaper gems, the Fossil Ivory, Prasitis, and Sapphirus; and which he describes as dry and composed, as it were, out of clotted blood, having also a variety of a paler tint. Pliny adds to this, that it was found both in Numidia and in Arabia; and as he, too, classes it amongst the *gems* (60), the Hæmatitis probably was no other than our Red Jasper.

But his Hæmatites, the *mineral* "lapis," as distinguished from the *gem* in another passage (xxxvi. 37), was a red oxide of iron, very serviceable in all diseases of the eye, and in the treatment of dysentery, and which Orpheus (650) fables to have owed its origin to the blood of Uranus which fell to the earth when he was mutilated by the scythe of Saturn. It continued to declare its true nature, for if steeped in water it again resolved itself into blood. It was a sovereign medicine for the eyes, *because* "that primeval god (Heaven) from whose veins it had flowed could not endure that mankind should, through loss of sight, be deprived of the view of his desirable countenance." Pliny's notice of the *gem* Hæmatitis is unmistakably a condensation of this part of the Orphic poem, although quoted as the book dedicated by Zachalias the Babylonian to king Mithridates; for that writer, besides the foregoing medicinal properties, "had extolled its mighty influence in ensuring success in petitions to princes and in all disputes:" the mystic poet having sung how "Dolon, aided by it, the gift of Helenus, had obtained the good graces both of Priam and of Hector;" and how "Ajax, had he not spurned the soothsayer's proffer of the stone, had certainly gained even Minerva's vote against her favourite Ulysses in the contest for the armour of Achilles." Pliny adds a bright yellow variety, "e fulvo candicans," the Indian name of which was *Menui*—an evidence that he is here speaking of a *gem:* although it is equally evident throughout the notice that he had no actual experience of either sort as then employed by Roman jewellers or engravers.

There was, however, very good reason why the later Romans should have become so fond of the Oriental Red Jasper as soon as it was introduced plentifully amongst them. The colour is a pure vermilion, both taking and well retaining the highest polish, in consequence of the fineness of its texture, sometimes traversed with parallel black hair-like lines; another, and a rarer variety, deepens into the richest crimson. The source of this fine stone has been lost: it was probably Arabia. Red Jasper is, indeed, still found abundantly in Sicily; but this kind is coarse-grained, and its tint is clouded, and verges more

L 2

upon a chocolate colour. The true antique *vermilion* is now only to be discovered in the specimens brought from Mexico. The paler sort, already mentioned as extensively employed by the early Egyptians for making charms, and exactly resembling unpolished Coral, was possibly the secondary Hæmatitis of Theophrastus, and, with more certitude, Pliny's Corallis.

3. A dark-green, opaque, close-grained Jasper, occasionally clouded with red, in great request with the Egyptian engravers for religious intagli from the epoch of the Pharaohs down to the Arabian Conquest. The Basilidan talismans affect this material above all others. It could have been no other than Pliny's *Molochites*, exhibiting as it does the peculiar dark-green of the mallow-leaf; that stone, too, being specially recommended by its virtue as an amulet. In fact, this Green Jasper appears to have been exclusively dedicated to the embodiment of Egyptian theosophic ideas (whether ancient or freshly developed upon a newly-imported Indian stock); for it is a rule without an exception, that no artistic works either in the Greek or Roman style ever present themselves upon such a material. A most rare variety (now only found in India) exhibits a pleasing opaque, sap-green : and on this sort is engraved the celebrated signet of Cleopatra (Marlborough), displaying her head, crowned with the phœnicopterus, in profile ; on the reverse, the bust of her patroness, Isis, within a shrine. The rarity of this beautiful kind is indicated by the importance of the subjects for which it has been selected by the ancients ; for another example (Bale) offers the portrait, deeply engraved in the Egyptian manner, of some African prince, wearing the peculiar, Chinese-like hat seen upon king Arcesilaus, of Cyrene, in the well-known Etruscan vase-painting, the "Silphium-merchants" (Berlin). Fragments may be seen

mingled with the other components of some Egyptian breccias; perhaps the actual source whence the old engravers extracted the gem. The material chiefly used for the Phœnician scarabei lately brought to light in Sardinia, though greatly resembling this in appearance, is of a totally different nature, yielding readily to the steel point, and being, perhaps, only a fine Serpentine. This Sardinian species has all the character of the *Smaragdite*, better known by its Italian name of " Verde di Corsica." Either of these would be aptly described by the comparison of their colour to verdigris, the way in which Theophrastus defines his " Prasitis."

4. A light-green, mottled with yellow after the manner of the Moss-agate, only quite opaque, and extremely hard, also serves as the frequent vehicle for Mithraic and Basilidan talismans. This would seem to be the Tree-agate of Orpheus (230), that insured plentiful crops if tied round the ploughman's arm, or to the horns of the oxen breaking up the field.

5. A bright yellow opaque stone, exclusively appropriated for talismanic engravings, astrological and Gnostic. This may represent Pliny's " Cerachates," of which Solinus observes, " that such as are of the colour of bees'-wax, being vulgarly plentiful, are despised." It is, however, also possible that he is here alluding to the common European Calcedony.

No stone held so high a rank in the alexipharmaca of both ancient and mediæval physicians as did the Jasper. Pliny and Epiphanius have been cited above in testimony to the virtues of the Grammatias. Even the sober and practical Galen declares ('Simp. Med.' ix.) " The Green Jasper benefits the chest and mouth of the stomach, if tied over them. Some people set it in a ring and engrave upon it a *serpent with radiated head*, just as King Nechepsos pre-

scribes it in his 13th Book.   Of this gem I have had ample
experience, having made a necklace out of such stones, and
hung it about the patient's neck, descending so low that
they could touch the mouth of the stomach; and they
proved of no less service than if they had been engraved
after the manner directed by King Nechepsos."   What
Galen understood by his "green Jasper" is placed beyond
all doubt by the still existing multitude of Plasmas belonging to his times, engraved with the serpent Chnuphis, often
accompanied by a Coptic invocation, the purport of which
is in one example made apparent by its Greek translation,
"that he would keep in health the chest of Proclus."   All
the amulets of this nature are confined to the Plasma and
greenish-yellow Calcedony.   Orpheus again (264) affirms,
"that if one offer sacrifice holding in his hand the polished
Jasper, colour of the Spring, he rejoiceth the heart of the
gods, and they shall satiate his thirsty land with showers,
for this stone bringeth down rain upon the parched-up
fields."   The transcendental virtues of the Red Jasper
were, as it would seem, sung by the same poet under its
appellative of Hæmatites, considered in the preceding
pages.   Pliny adds the belief that the vermilion-coloured
Achates rendered athletes invincible, a notion that may
account for the frequency of the engraving of Hercules
wrestling with the Nemean lion upon that sort.

Supported thus by the authority of Galen, the efficacy
of the Jasper continued undisputed far down into the
seventeenth century.   De Boot testifies that in his own
practice he had observed effects, scarcely credible, following from the application of the Red in cases of hemor-
rhage, whether internal or caused by wounds; and mentions
the prevailing belief that the Green, if engraved with the
figure of a scorpion when the Sun was entering that Sign
of the Zodiac, was a sure preservative against the stone in

the bladder. His patron, the Kaiser, so highly admired
the whimsical variety of shades and colours wherewith
Nature has painted the different species, that he caused a
table to be made up out of them, so ingeniously arranged
that their contrast produced a landscape, with its mountains, rivers, trees, and clouds. The work took several
years to complete, and cost "many thousand gulden." In
this piece of "lavoro di commesso" we discover the first
germ of the afterwards so celebrated Florentine manufacture, the "Pietra dura" mosaic, that constructs true
pictures of bouquets, &c., imitating the flowers with the
natural colours of slices of Agates and Jaspers cut to the
shape required: an art involving incredible labour, and
corresponding costliness in its finished productions.

LAPIS LYDIUS: *Touchstone: Assaying: Alloys: Foils.*

BACCHYLIDES, the lyric poet, who flourished B.C. 450, introduces in a simile,* and Theophrastus minutely describes this most ancient of all methods of assaying the precious metals (47): " Wonderful again is the property of that stone used for testing gold, for it appears in fact to possess the same power as fire, for that also acts as a test. For which reason some are puzzled about it in consequence of their not viewing the question in a proper light. For the stone doe's not really try the gold in the same manner as the fire; because the latter acts as a test by changing and altering the colour of the metal, whereas the former tests it by friction. For it has the power, as it appears, to distinguish the nature of each metal, the gold and the alloy; and it is said that lately a much better kind of Touchstone than the old one has been discovered, capable of testing not only gold after the refining, but also copper containing gold, as well as silver (similarly auriferous), and of showing how much alloy has been mixed with the standard of the gold coin. The assays are taken from the smallest quantity, for the least weight is a grain of barley, the next the collybus (1¼ gr.), then the quarter, then the half-obol (5 gr.): from which they ascertain the amount of the alloy—all such Touchstones are found in the river Tmolus. Their form is

* Λυδία λίθος μανύει χρυσὸν ἀνδρῶν δ' ἀρέταν σοφία τε παγκρατής τ ἐλέγχει ἀλήθεια. "The Lydian stone tests gold, but wisdom and all-powerful truth prove the virtue of man."

smooth, and like that of a counter, flat, not spherical, but their size about twice that of the biggest counter used. For the purpose of assaying, there is a difference between the upper part of the stone which has been exposed to the sun, and the underside, the latter being the better test. This is a necessary consequence, inasmuch as the upper side is of a drier nature : for moisture is an impediment to its taking off the metal; hence in very hot weather the stone does not act equally well, for it then emits a certain dew out of itself, by reason whereof it becomes slippery."

Pliny (xxxiii. 43) refers to this description by Theophrastus, adding that the Touchstone, besides its Latin name "coticula," was called the Lydian stone, and the Herculean. He further states, what seems incredible, that experienced assayers, by rubbing off a piece of any ore on this stone, "cum e vena ut lima rapuerunt experimentum," were enabled to declare without mistake how much gold, silver, or copper it contained, within one scruple, "scriptulari differentia." From its use the stone received the epithet of "Paragonius" or comparative, whence by a singular transition of ideas comes our 'Paragon' in the sense of 'Pattern of excellence.'

The present Touchstone is a black Jasper of a somewhat coarse grain, and the best pieces come from India. The Italian goldsmiths employ it in the following ingenious manner. They have a set, strung on a ring, of 24 "needles," little bars of gold, each of a known and marked standard from one carat up to twenty-four (or fine). Taking the gold to be assayed, they rub it on the stone; by the side of the streak it leaves, they rub the needle which seems to the eye the nearest in quality. Next they pour "aqua regia" on the two streaks, and if the solvent produces the same effect on each, it proves that the gold in the piece and in the needle is of the same standard. If there be a difference perceptible, they try another needle, until an exact coincidence is obtained. It is very singular that Tavernier should have known nothing of this method before he saw it generally employed by the *Banyan* money-changers: for as the Italian craftsmen are as immutable in their processes as the Hindoos themselves, they must have possessed this simple method of assaying in that traveller's days.* To exhibit the economy of the Banyans he adds that they carefully wipe the stone, after using it, with a lump of wax, which thus in course of time becomes charged with an appreciable quantity of fine gold.

Exactly a hundred years before, Abul-Fazel writes: "In this country the *serafs* (money-changers) know the degree of fineness from the colour and brightness of the metal, but for the satisfaction of others this grand rule has been introduced: the *bunwary* is composed of a number of bars of copper, or such-like metal, on the point of every one of which is fixed a small piece of gold and the degree of purity written thereon. When they want to assay the newly-imported gold they draw on the Touchstone a line of this and a line of the bunwary: and by comparing them together they discover the degree it is of." (Ayeen Akbary, p. 8.)

The assay of gold by fire has been described under *Obryza* (AURUM). The Roman assay of silver was equally simple. Filings of the article to be tested were thrown upon an iron chafing-dish made white-hot: if the silver kept its colour it was proved *fine;* if it turned red

---

* A truth curiously exemplified by Castellani's own experience: " But it was only in a remote corner of the Umbrian Marches at S. Angelo in Vado, a little district hidden in the recesses of the Apennines, far from every centre of civilization, that we found still in use some of the processes employed by the Etruscans. There yet exists, in fact, in this district of Italy, a special school of jewelry traditional, somewhat similar, not indeed in taste or elegance of design, but at least in method and workmanship to the ancient art."

(containing copper), of the second quality; but if black (betraying the presence of lead), it was pronounced base. Pliny exposes an ingenious fraud of the silversmiths intended to baffle this test. They kept the chafing-dish beforehand steeped in urine, the salts of which, when the iron was heated, fixed upon the filings assayed, and, whitening them for a time, so enabled the piece to pass muster for fine standard. Another and more ready test was to breathe upon the polished surface, the more quickly the breath dispersed, the better the quality of the plate. As the silver coin even in Pliny's age was falling rapidly in its standard, we may suspect that the plate also was of a quality more or less debased, according to the honesty of the silversmith.

It is a curious fact that the Chinese *Tutenague* or " white copper " (of late years so largely manufactured here with the name of German silver) had already, under the Ptolemies, found its way to Europe as a substitute for the precious metal. Crinagoras sends to a friend for a birthday present, " a *copper* flaggon (*olpe*) exactly resembling silver, an Indian work," together with a neat epigram. (Anth. vi. 261.)

The famed " Corinthian Brass " may be classed indifferently amongst the ancient alloys of gold or of silver. Pliny describes three qualities of this composite metal, which was an alloy of gold, silver, and copper. In fact, it was much the same as what our jewellers dignify with the name of gold, and employ for all articles not Hall-marked for 18 carats fine. The first quality was yellow-coloured, in which the gold preponderated; the second resembled silver, that metal constituting the greater quantum of its mass; the third contained all three in equal proportions. Such alloys would all now pass for gold, and indeed the first is better than our 12-carat gold, and the third equal to our 8-carat, so much employed in neck and guard chains, and called by the sellers *fine*: but the Romans disdained to give the title of "aurum" to the noble metal unless totally separated from all baser admixture. This true "Corinthian Brass" was only used for making dishes and drinking-vessels. The wealthy Spurinna, a friend of the younger Pliny's, boasted of a complete dinner service in this rare material. Pliny (xxxiv. 3) laughs at those dilettanti who gave that name to the metal of the old Greek statues they so eagerly collected; adding, that "they pretended to a knowledge in this matter merely to distinguish themselves from other people, without in reality having any deeper understanding of it."* This he proves by showing (in his catalogue of statuaries) that the great masters whose material these connoisseurs styled promiscuously "Corinthian," had flourished all of them some ages before the fall of Corinth (B.C. 147). For he accepts as true the account given of the accidental discovery of this alloy in consequence of a fire, during the sack of that city, having melted a number of works in the different metals that had been brought together (by the plunderers). Of the same tale Petronius makes his hero Trimalchio, the wealthy "snob," give a comical version: "But that you may not take me for a know-nothing, I understand quite well how the Corinthian Brass first came about. At the sack of Troy Hannibal, a cunning fellow and a big rogue, heaped up all the gold, silver, and bronze statues into one vast heap, and then set fire to it. The metals fused and

---

* Martial lets us know a curious test of the genuineness of the composition : his virtuoso

"Consuluit nares an *olerent* vasa Corinthum,"

and it is certain that the above-named admixture of the precious metals would go far to neutralize the peculiar nauseous smell of the baser one in the resultant.

ran all together. From this the workmen took and made pots, dishes, and statues. So arose the Corinthian metal; one thing out of several, but neither this nor that." But the truth that the mixed metal in question had become famous * more than a century before the catastrophe of the city giving it the name, is fully established by the mention of the same as a precious article by Hippolochus, who was a pupil of our Theophrastus. In his most amusing description of the wedding of Caranus, a Macedonian grandee, the first course consists of a bronze dish of the *Corinthian composition* (κατασκευασμάτων) set before each one of the twenty guests. This dish was covered with a flat cake of equal extent, whereupon were piled fowls, ducks, pigeons, partridges, and a goose, an assemblage attesting its ample dimensions; both dish and contents being a present to be carried home untouched. The value of the material is indicated by the fact that it ranked with the other pieces of plate bestowed upon his astonished friends by the extravagant generosity of the bridegroom. These were a silver dish containing hares and similar game, another with a large roasted pig stuffed with small birds, beccaficoes, &c., and finally a square *pinax* supporting a boar roasted whole upon a silver spit. (Athen iv. 128.)

By a very remarkable coincidence the Chinese have in our day the same composite metal, the same legend to account for its origin, and the same enthusiasm for collecting specimens of it as the Romans of old. Between the years 1426-36, in the reign of the Emperor Siouan-Tsoung, the imperial palace took fire, and the violence of the conflagration, which lasted several days, was so great that immense quantities of gold, silver, and brass, were melted by the

---

* Something of the kind must have been the "*fine copper*, precious as gold," two vessels whereof are named in the list of Artaxerxes' donaria to the Second Temple. (Ezra, viii.)

flames, and all these metals were mingled together. The
Chinese annals add that a number of vases were made at
the time out of the material obtained by this commixture.
These vases are much esteemed in China, and fetch a high
price. (Du Halde.)

I have never been able to discover if any vessels in this
composition yet exist; but their non-appearance would be
no argument against the antique, and large, employment
of the alloy. The article *Cælatura* has shown how few of
the innumerable pieces of gold and silver plate of those
times have come down to us, and the precious alloy of the
Corinthian would have consigned it to the same crucible
as the works in the purer metals. Add to which, its
surface *now* would (from the mixture of copper) be so
coated with *patina* as to render the metal undistinguishable
to the eye from ordinary bronze. Perhaps therefore it
may be the actual material of the magnificent discus (mirror-
case) found at Paramithia, embossed with figures in extremely high relief, with draperies of silver, representing
the Visit of Venus to Anchises. The numismatists of the
Revival were fond of espying the true Corinthian in the
material of the large Roman coins, misled by its bright golden
colour: they never reflected upon the trifling value of such
pieces (four to the denarius) when actually in circulation.

That the alloy called *aurichalcum* was but another name
for the *Pyropus* (as designating a *metal*) may be inferred
from its name "gold-copper," and from the manner in
which Pliny alludes to its appearance and use. He does
not anywhere give its composition, which of itself is a
reason for supposing he had already described it under
another trade-name.* He quotes the best bronze, the

---

* The *Pyropus*, copper with one-fourth of gold, only used for making
foil "pretenui bractea ignescit" (xxxiv. 20). There was also a leaf-
copper (our Dutch leaf) lacquered with ox-gall, called *Coronarium*,
from its use in theatrical crowns.

Livianum, as "imitating the excellence of aurichalcum" (xxxiv. 2), in its fine colour evidently. The latter quality is further indicated by the fact he mentions that aurichalcum was used for foiling the paler sorts of the Chrysolithus. That it was a costly metal is also clear from Horace's (A. P. 202) allusion to its use for binding the large and expensive flutes then in fashion at the theatre : besides Ovid's introduction of the same article into the architecture of heaven.

> " Regia Solis erat sublimibus alta columnis,
>   Clara micante auro flammasque imitante *pyropo.*"

Its manufacture was preserved by the Italians down to the Revival, but only as the material of *foils* for precious stones ; as will appear from a consideration of the recipes for making the different foils given by Cellini in his ' Orefeceria.' Pliny only mentions three kinds of them, the aurichalcum, the pyropus, and a silver foil (argentæ bracteæ) for the Sard. Now Cellini directs the jeweller how to make thin sheets out of gold, silver, and pure copper, united in varying proportions according to the *colour* required. This last was produced by holding the polished sheet of metal (hence its name from *foglio*) over a charcoal fire until it acquired the proper tint: exactly as they now-a-days *blue* steel.* It will be perceived that Cellini's alloys answer to the various kinds of Corinthian Brass. Foils in the present trade are manufactured much more simply, by painting a sheet of silver-foil with the required colour mixed in a transparent varnish.

---

* Cellini says nothing of the curious recipes given by Porta for producing the different colours in the metallic sheet, by means of the *smoke* of different substances burnt on the furnace below : the sapphire-colour by the smoke of the blue feathers of a drake, the emerald by box-leaves, the ruby by scraps of scarlet cloth, &c.

LYNCURIUM : Λυγκούριον : *Jacinth.*

THE Lyncurium derives its name from the strange notion that it was the urine of the Lynx petrified, and it is so described by Theophrastus (28): "Equally wonderful is the Lyncurium (for out of this also signets are engraved, since it is very hard, exactly like a real stone), for it attracts in the same manner as Amber, some say not only straws and bits of wood, but even copper and iron, if they be in thin pieces, as Diocles also hath observed. It is highly transparent and *cold* to the touch. That produced by the male lynx is better than that by the female, and that of the wild lynx better than that of the tame, in consequence both of the difference of their food, and of the former having plenty of exercise, the latter none—hence the secretions of the wild are the more limpid. Those practised in the search find it by digging; for the animal endeavours to conceal the deposit by scraping up the earth after having voided it. There is a peculiar and tedious method of working up this substance also (as well as the Smaragdus)." Ovid repeats the same story (Met. xv. 413).

> " Victa racemifero lyncas dedit India Baccho
> E quibus ut memorant quicquid vesica remisit
> Vertitur in lapides et congelat aëre tacto."

Pliny indeed (xxxvii. 13), after apologising for mentioning the Lyncurium, "to which he was forced by the pertinacity of writers on mineralogy," quotes the above passage from Theophrastus; and declares that unless the thing were

Amber,* it nowhere existed at all in his time, and that the whole tale was a fiction. But Theophrastus knew Amber too well to make so ignorant a mistake, in fact he proceeds to describe it accurately enough in the next chapter, stating its most obvious properties, and informing us that the Greeks got it from Liguria, *i. e.* from the Celts the nearest to them. Besides, the *coldness* of the substance in his *Lyncurium*, and its being used for intagli, prove that Theophrastus is describing an amber-coloured *stone:* for real Amber is remarkably *warm* to the touch, and much too soft and brittle to be used for engraving on. Pliny rejected the account without due examination, partly disgusted by the ridiculous story of its production; partly because the stone known by that name to Theophrastus, was then reckoned amongst the Chrysolithi by the Romans.

Epiphanius (*Ligurius*) suspected that the Λιγούριον of the Septuagint, of which he could find no account in any Greek mineralogist, was the Hyacinthus of his own times, on the ground that so important a stone could not have been omitted by Moses; and here we find the first germ of the subsequent confusion of two very different things. Jerome writing to Flavilla makes the same remark and in almost the same words. Isidorus, however, correctly explains "Ligurius" as synonymous with Lyncurium, abbreviating the words of Theophrastus.

There can be no doubt that the *gem* described by the latter author is our *Jacinth* (Zircon) the yellow Jargoon,†

* And to a certain extent he is right; the entire fable springing from the confusion between *Ligurius, i. e.* Ligurian stone, and Lyncurius, their difference in the Greek pronunciation merely depending upon the duplication of the γ. Now we find that Amber was first obtained by the Greeks from these Cisalpine Gauls, and naturally, therefore, would be called the Ligurian gem.

† It is satisfactory to find this explanation confirmed by the authority of Corsi, who gives 'Giacinto Ambrato' as the synonym for 'Lyncurius.'

(G) M

distinguished by having for its chemical base the˙earth Zirconia, peculiar to this family. This exactly resembles Amber in colour, refraction, electricity and levity, and the sole distinction is its excessive hardness. We find in it also the two kinds mentioned by the Greek naturalist: a dark orange, extremely agreeable in tint (the male); and a pale yellow of extraordinary lustre (the female).

Another argument for their identity is the frequent employment of the Jacinth by the Greeks, for intagli in early times, and by the Romans for camei also. But for the latter purpose they preferred the darker kind, which, thus worked out, is extremely effective. The style of antique engravings in this gem is altogether peculiar, so as to be easily recognised even in the impression from such an intaglio. It is characterized by a certain fluidity (*lavoro bagnato*) and roundness of all the lines, and a shallowness of engraving adopted apparently to avoid all risk of fracture in working so porous a stone. This porousness is manifest even to the naked eye, for a Jacinth held up against a strong light appears like a mass of petrified honey, or guava-jelly. This difficulty in the engraving is remarked by Theophrastus in the passage above quoted, such at least appears to be the meaning of the obscure expression, " γίνεται δὲ καὶ κατεργασία τις αὐτοῦ πλείων." He alludes here to what he had just before said of the Smaragdus: " there is a particular mode of working this stone so as to give it lustre, for in its native state it is not brilliant." From the similarity of texture in the two there is reason to infer that he means by this peculiar mode that use of the Marcasite (instead of the usual Emery) which Ben Mansur describes as the only possible method for polishing the Laal (Spinel), one variety of which he gives as yellow, another as green. For, as far as regards the peculiar lustre, texture, and electricity, as well as

hardness, the antique Jacinth presents a wonderful analogy
to the brown Spinel; only differing in this, that the former
shows a rich orange, occasionally tinged with crimson, the
latter a crimson tinged with blue, or brown.
This statement explains the peculiar style of the Greek
intagli in Jacinth, worked out as they are in flowing and
shallow hollows, evidently by a *technique* dissimilar to that
employed in the Sards and Agates of the same period.
From its porous texture, in spite of its great hardness (7·5,
that of the Spinel being 8), antique works in it have always
a worn and scratched surface; and this so invariably,
that a Jacinth exhibiting a perfectly polished exterior may
justly be suspected of being a modern work, or at best a
retouched antique. Even the interior of the design, unless
where protected by unusual depth of cutting, will be
found to have suffered in a conspicuous manner from the
effects of friction and of time.
From its high electricity the soft wax (such as the
Romans used in sealing) * adheres to the intaglio, and will
hardly afford a good impression: it is therefore singular
that so many fine intagli should have been executed in it.
But the Greeks, instead of wax, used pipeclay † (the *creta*

* And the only substance used for the purpose until about two centuries ago, when the Dutch introduced from India the hard sealing-wax,
or rather, as the Germans more properly call it, *siegel-lac.* The old composition is still retained amongst sculptors for modelling in; it is coloured
bees'-wax, rendered plastic by the mixture of a few drops of turpentine, readily yielding to the warmth of the finger. Falstaff says, "I
have him tempering between my finger and thumb, and will seal with
him anon."

† From its use, the Greeks called this γῆ σημαντρὶς, "sealing-earth."
Athenæus quotes amongst the witty sayings of Lais, that once, on re-
ceiving in this material the impression of her lover's signet, in token
that she was to come to him, she replied to the messenger, "I cannot,
πηλὸς ἐστί," which would bear the meaning that it was too dirty for
her to venture out. On the same usage is based a wild legend; the
Erythræan Sibyl having obtained from Apollo the gift of immortality

M 2

of the Romans, ancient and modern), for their seals, to which the same objection does not apply. One of the finest intagli in Jacinth known is the head, formerly considered that of Pompey, but now with more justice attributed to Mæcenas, (Rhodes), which is rendered still more valuable by the supposed signature of the engraver Apollonius.

The deep-coloured sort, of the richest orange-brown, usually found with antique camei, for the most part heads of Fauns, Masks, and such like Bacchic ideals, carved out of it, has a better claim than the dark Carbuncle to be considered the *Morio** (so called from its mulberry colour), which Pliny notes was employed for engravings in relief, "ad cetypas scalpturas aptantur." As he makes such an observation concerning no other gem, it may be concluded that the Morio was then in chief request for that purpose. This attribution is also supported by his definition of its three varieties; the Indian, entirely dark yet translucent, called also Pramnion (after a very strong red wine): the Alexandrian, with which a Carbuncle tinge was mingled;† and the Cyprian, verging upon the colour of the Sard. Köhler ('Ueber den Sard,' &c.) maintains that by Morio the

provided she never again beheld her native earth, departed and settled at Cumæ. But at last, worn out with old age, and thoroughly weary of life, she applied to the senate of her country, who in pity returned her a letter sealed in the usual manner: and no sooner did the Sybil set eyes upon the seal made of her natal soil, than the spell was broken, and she delivered from the bondage of existence.

* Or Mormorio (Jan's reading); a name the seeming connection of which with μορμώ may account for its selection by the ancients for suchlike bugbear images as Satyric masks and Fauns' heads.   ·

† The variety now appropriately termed by the Italians "Giacinto-Guarnaccino" (from *vernaccia,* a white wine), in whose orange the red tint strongly predominates over the yellow. Under this name the brownish-red Spinel is frequently confounded: it being almost impossible to distinguish one from the other, except by the difference in their relative specific gravity.

Romans understood the Black Sard (our Black Agate), only red and translucent by transmitted light; but though μορμὼ masks, in intaglio, are frequent enough in this variety, those in *relief* never occur in it. De Boot, with as little foundation, supposes it Obsidian.

The exact nature of the Mormorio is much elucidated by the discovery made (1838) in a tomb near Kertch, by Alexis Tscherkow, where lay a female skeleton, richly adorned with her choicest jewels—a laurel crown, in massy gold, on her head; earrings, with pendants, formed as Bacchic genii; and two necklaces, with hollow gold beads: " crepundia," as the Romans called them. On her fingers were eight rings, of which three had been votive, to judge from their extraordinary and unwieldy size. One was set with a bust of Pallas, in three-quarter face, her helm crested by a bird with drooping wings (not her proper owl, however), in high relief, of the finest Greek work: cut on "une pierre qui ressemble au grénat le plus foncé." The second bore the same relief, but with the head in front face. Both these stones are nearly two inches long—their form, a wide oval. The third had a superb Siriam Garnet, nearly as large, with an intaglio, a god leaning on a sceptre, and placing his hand on the head of a smaller figure standing on a basis. (Published in the ' Bull. de l'Inst. Arch. Rom.' 1840).

A perfect Jacinth is indeed a splendid ornament to the finger; much more agreeable and superior in tint to the best Brazilian Topaz, having a peculiar golden lustre mixed with its rich orange; however, it is now completely out of fashion, and consequently of but little selling value, such are the unaccountable, unreasoning, mutations of taste in these matters.

Our Jacinth was not the *Lychnis* of Pliny, as some have taken it to be, the latter was indeed electric and hard to engrave; but the definition of its two distinctive colours, as purple and crimson, (purpura-coccus), proves it indubitably the Spinel ('Precious Stones,' p. 228). The former was reckoned by the Romans (at least the yellow kind) amongst the Chrysolithi: a numerous genus, including every yellow stone, from the Oriental Topaz to the humble yellow Crystal, or Cairngorum. Some of the Indian Chrysolithi were evidently Jacinths, and the notice of the defect in the Arabian that they were cloudy and filled up as it were with their own filings or dust, aptly expresses the peculiar porousness of this stone.

Modern jewellers universally term Jacinth, a gem belonging to a totally different species, the Cinnamon-stone (Essonite), a reddish brown Garnet, greatly resembling it in outward appearance. But the Cinnamon-stone may readily be detected by its total want of electricity, and also by the clearness or glassy nature of its substance when held against the sun, so different from the porous Jacinth. It possibly was Pliny's (44) Chryselectrus, "dyed with saffron, and only to be distinguished from a paste by its coldness." The Cinnamon-stone was engraved upon by the Romans and the later Persians, but not to nearly the same extent as the real Jacinth. The Renaissance artists, however, employed both very largely for works in relief, principally the pale yellow kind, which alone was then at their command. The ancients very rarely cut their camei in single-coloured stones, except in the dark-red or the purple Morios fancied by them appropriate to the subjects.

The derivation of the name "Jacinth" is curious, and the following appear to have been the steps by which it came to be transferred to the modern gem from one of a totally different family and appearance, which thereby has in its turn lost its original designation. "Jacinth," (the French retain the original "Hyacinthe,") comes to us from

the Italian "Giacinto," formed according to the usual rule
of that language from the Latin "Hyacinthus."* Although
it is perfectly made out that the Hyacinthus of the earlier
writers, Pliny and Solinus, was our *Sapphire*, and that its
distinctive quality to them was its azure colour, yet we
find Epiphanius, at the close of the fourth century, describing
under that name all the three varieties of the Precious
Corundum—the Blue, the Red, and the Yellow; following
the Indians, who give the same generic title to all, but
distinguish each sort by an epithet, denoting its colour; or
rather the Persians, who divide the "Jacut" (whence
"Hyacinthus") into six classes. Now *Sapphirinus*, ultramarine, being employed to indicate the Blue, or most
esteemed sort,† *Rubinus*, rosy, the next; these epithets, the
noun being dropped, were used absolutely for designating
those two varieties, and thus became fixed in the jewellers'
language as *Sapphire* and *Ruby*. But a third sort remained,
the least valuable of all, the *Citrinus* or lemon-coloured,
known by that name even as early as when Marbodus
wrote. This last sort being the most common retained
the generic name, and continued the *Hyacinthus*. After the
importation of gems fell off and gradually became extinct,
all yellow stones of superior hardness then circulating
throughout Europe got confounded together, and the more
readily because nothing being now required in the way of
engraving or even of polishing these relics of ancient
prosperity, the eye was the sole means of discriminating
their quality and value. Now, the Zircon, a gem the most

* The corruption goes back to an early date: in Pope Gerbert's
fine hymn "Extra portam," composed in the eleventh century, we already
find it :
  "Sed quæ gemma muros pingit,
    Quis Chalcedon, quis *Jacinthus*. . .

† "Albertus (Magnus) tamen ponit *Sapphirinum Jacinthum* princi-
patum obtinere." (Spec. Lap. p. 112.)

abundantly employed by the ancients both for engraving on and for ornament, was therefore the most plentiful amongst the *yellow* gems treasured by the barbarian plunderers: it was hard and electric; it came from India only, the source of the Corundum species as well, and thus by degrees it usurped and engrossed the name of Hyacinthus, previously borne by the rarer yellow gem, which now became the *Citrinus.* Besides, it would require an experienced mineralogist to distinguish by the eye alone a pale bright Zircon, or Essonite, from the Oriental Topaz.

Pliny (xxxvii. 13) puts down as equally false with the story of its formation, the notion that the Lyncurium, if drunk in wine, or even worn, would expel the stone in the bladder and cure the *jaundice*—an early allusion this to the "Doctrine of Signatures," *i. e.* that each substance bore a natural mark (in this case, the colour), pointing out the malady for which it was a specific.* Marbodus prescribes this stone for complaints of the chest, for the jaundice, and the diarrhœa. But his contemporary Psellus ('De Lapid.' Preface) reckons the Lyncurium amongst the stones about which nothing was then definitely known.

* A doctrine yet more plainly enounced in his description of the Icterias, "resembling in colour the human skin when of a sickly yellow, and *for that reason* esteemed efficacious against the jaundice."

MAGNES: Μάγνης: *Loadstone.*

THE Loadstone, or Magnetic Iron-ore, is a compound of Protoxide and Peroxide of Iron. It is black, heavy, and compact; sufficiently hard to take a good polish, which gives the surface a metallic lustre, like iron polished with black lead. Pliny (xxxvi. 25), on the authority of Nicander, states that the Magnet took its name from the herdsman who first discovered it in Mount Ida, by its attracting the nails in his soles, and the ferule of his staff, as he walked over the bed. Pliny notices the great abundance of it in Spain; in fact, the richest iron-mines at present worked, those of Elba and Sweden, consist entirely of Magnetic ore.

Sotacus divided it into five species: the Ethiopian, the best, and sold for its weight in silver; the second, found in Magnesia of Macedonia; the third, in Hyettus of Bœotia, redder in appearance than the second; the fourth, in Alexandria Troas; the fifth, in Magnesia of Asia, which was the worst of all, being white and like a pumice-stone.* That of the Troas was black and but feebly attractive, and was therefore considered the female of the species. It had been observed that the blacker the colour (*i. e.* possessing

---

* Which two qualities are a sufficient proof that this mineral was no Magnet at all, but merely had been confounded with it by unpractical describers in virtue of the similarity of the name. It was a stone of which Theophrastus (quoted below) has taken notice.

more of a metallic, steely lustre) the more powerful the Magnet.

It is a singular omission on the part of Theophrastus that he should make no mention of this mineral, though described by his master Aristotle accurately enough; his *Magnetis* would appear to be some Talcose rock from his notice (41): "Some stones admit of being sawn, others of being carved, or turned in the lathe like the stone *Magnetis*, which holds a high rank for its beauty, and is by many much admired on account of its resemblance to silver, though it really has no affinity to that metal."

The singular nature of the Loadstone excited the wonder of the ancient naturalists. Pliny exclaims, " What is more inert than the rigidity of a stone? Yet lo! Nature hath given it feeling and hands. What more contumacious than the hardness of Iron? She hath bestowed upon it both feet and complaisance: it is attracted by the Loadstone; and that metal, the subduer of all things, runs to meet something or other that is bodyless (inane), and when it comes near leaps towards it, is held by it, and clasps it in its embraces. Hence is it termed Sideritis; and by others, for the same reason, *Heraclios*, or the stone of Hercules." Claudian has an elegant poem (translated at the end of this article) upon a shrine, containing a statue of Venus made of Loadstone, and another of Mars in iron. When the festival of their marriage was celebrated these statues were brought near, and by mutual attraction appeared to fly into each other's arms. There can be no doubt the poet is describing what he had himself witnessed.

Dinochares (or Timochares) the architect had, according to Pliny (xxxiv. 42), commenced vaulting over the temple of Arsinoe, at Alexandria, with blocks of Loadstone, with the idea that her statue, to be made of iron, would thus remain suspended in mid-air. This tradition was in all

likelihood the source of the mediæval fable concerning
Mahomet's coffin thus balanced in his monument at Mecca.
The Romans were acquainted with the fact that the Loadstone communicates its virtues to iron; they amused themselves by putting together a series of magnetized iron rings,
so as to form a long chain by their cohesion. The metal
thus magnetized was called by the ignorant vulgar " Quick
Iron," and supposed to inflict more dangerous wounds.

The *Gold Magnet* of Ben Mansur is evidently nothing
more than the Magnetic Iron-Pyrites: its property of
attracting gold being a mere theoretical inference based
upon its colour. The same holds for his Silver Magnet.
His *Camahen* or *Ass-stone*, "very hard, but which breaks
up into splinters," is clearly our Magnetic Iron-Ore, called
improperly Black Hematite. His epithet exactly applies
to its grey colour. Now, taking into account the general
use of this substance in ancient *Persia* for intagli (in his
age universally regarded as talismans), the word *Camahen*
as denoting an engraved stone of special importance, and
its introduction in that sense into Europe by the Crusaders,
furnish a plausible explanation of the origin of the much-
disputed term, Cameo. (See CAMEO.)

The earliest, also the latest, essays of the Glyptic art
amongst the ancients were made upon the Loadstone. It
is the favourite material for the Assyrian cylinders, and
equally so for the Cuphic signets, which close the history
of Oriental intagli. So common a material was disdained
by the Greek engravers, and after their example by the
Romans, until the diffusion of Eastern doctrines in the third
century brought the substance again into favour as an
amulet, and the Gnostics (at least those sects whose tenets
had rather a Persian than an Alexandrian root) used it
very largely in the manufacture of their talismans. As
may be well supposed, nothing valuable in the artistic

point of view is ever to be found in this stone : the only tolerable intaglio in Hematite that has come under my notice being a bust of Abundantia, in the Marlborough Cabinet. Yet its texture is compact, takes a high polish of a steely lustre agreeable to the eye, which in spite of its comparative softness it retains unimpaired by time.

The name *Sideritis* was also employed to distinguish one species of the Adamas: the electric property of the one being confounded with the attractive of the other. Similarly we find the name *Androdamas* equally applied to both species. The Adamas (whether the Diamond or Pale Sapphire), in virtue of its superior value, was believed not merely to greatly surpass the Loadstone in attractive force, but even by its presence entirely to deprive the other of its natural powers. (' Precious Stones,' p. 102.)

The Hæmatites, included by Pliny under the head of "Magnes," was the reddish-brown compact iron-ore still so-called, and used in the burnishing of metals, especially for smoothing down the surface of the gold-leaf in gilding and damascening. The best kind was procured by the ancients from Zmiri, in Ethiopia ; at present Compostella supplies all the jewellers of Europe. The only use the ancients made of this mineral was, burnt and powdered, in eye-salves : its astringent quality being doubtless of some efficacy in relieving inflammation of that organ. A curious but fanciful test of the true Hæmatites lay in its attractive power being restricted to the common Magnet, and not acting upon any other metal.

Orpheus extols (302) the virtue of the Loadstone as conciliating to the wearer the love both of gods and men ;* and prescribes it as a sure test for ascertaining if a wife has been faithful to her lord during his absence, for if put

---

* A belief indicated by the subject, the Three Graces, very frequently to be seen engraved upon this stone.

under her pillow as she sleeps, she will, if an adulteress, be dashed out of bed by its influence.

> "If e'er thou wish thy spouse's truth to prove,
> If pure she's kept her from adulterous love,
> Within thy bed unseen this stone bestow,
> Muttering a soothing spell in whispers low;
> Though wrapp'd in slumbers sound, if pure and chaste,
> She'll seek to strain thee to her loving breast:
> But if polluted by adultery found,
> Hurl'd from the couch she tumbles on the ground."—Ver. 312.

Marbodus adds another more mischievous (if possible) quality: that its powder strewn secretly upon the embers will drive all the inmates out of a house panic-struck, and give the operator an opportunity to rob it unmolested.

> "If a sly thief slip through the palace door,
> And strew unseen hot embers on the floor,
> Then powder'd loadstone on these embers spread,
> The inmates flee, possess'd with sudden dread.
> Distraught with horrid fear of death they fly,
> Whilst from the square the vapour mounts on high,
> They fly: within the house no soul remains,
> And copious spoils repay the robber's pains."—Ver. 305.*

Orpheus (636) sings, moreover, at great length the mystic virtues of the Hæmatites; not, however, the magnetic ore just noticed, but a soluble oxide of iron, of a styptic quality, and an excellent medicine for the eyes (JASPIS, p. 147.)

The Loadstone was used in glass-making, being supposed " to attract to itself the liquid of the glass in the same way as it does iron" (xxxvi. 66). Probably it acted as a flux in promoting the union of the silica with the soda. As the green and blue tinge of common glass is due to the small

---

* The Hindoos attribute to this stone the power of maintaining eternal youth in the human body. Garcias mentions an *aged* king of Zeilan who had all his dinner-service made thereof, in this belief, as the lapidary, the maker of the dishes, had informed him

proportion of iron naturally existing in the unpurified ingredients used in the manufacture, the remarkably deep blue and green characterising Roman glass to such a degree as to serve for a means of immediately distinguishing it from the production of later times (when fragments of the various periods are discovered jumbled together in the ruins of long-inhabited sites), may be due to the quantity of the combined oxides of that metal thus purposely introduced.

. It is, however, likely enough that Pliny here is alluding to the then recent discovery of colourless crystal-glass, and, confusing the Loadstone with a totally distinct mineral bearing the same designation of its locality, *Magnes*, may in reality mean the oxide of Manganese, which from its property of purifying the *pot-metal* bears the popular name of *glass soap.* The word Manganese comes from the old chemical appellation given to this mineral from such a use, *Maganesia Vitrariorum*, itself evidently a form of *Magnes.* This mineral was used by the Etruscans for making the black varnish covering their vases.

CLAUDIAN. Ib. V.

" A stone there is which people Magnet style,
 Dull, dark in colour, of appearance vile :
 Unlike to such as deck the comb'd-back hair
 Of princes, or the neck of maiden fair ;
 Or such as on the golden buckles shine,
 Which by their clasp the imperial belt confine.
 Yet such its wondrous force, it far outweighs
 All beauteous ornaments, all jewels' blaze,
 Or all those treasures which on Eastern shores
 The Indian 'midst groves of coral red explores.
 From steel it draws its force—from steel it lives ;
 'Tis this its food, 'tis this its banquet gives,
 And hence renews its strength ; pour'd through its veins,
 The rugged aliment its life maintains.*

---

* The Italians still keep the loadstone immersed in iron filings, "per mantenerne la forza."

Of this deprived, its frame exhausted lies,
Fierce hunger gnaws, and thirst consuming dries.
 With gilded ceilings deck'd a temple shines,
And two Immortals grace two common shrines,
Mars, scourging nations with his bloodstain'd spear;
And Venus, solace sweet of human care.
Different their mould—in iron Mars commands;
Sculptured in Magnet lovely Venus stands.
Their nuptials high with solemn rites to grace
Prepares the priest, the guardian of the place.
The blazing flambeaux lead the dancing quire;
High o'er the gates the myrtle boughs aspire;
With heap'd-up roses swells the marriage-bed;
The nuptial chamber is with purple spread.
Behold a marvel! Instant to her arms
Her eager husband Cythereia charms;
And ever mindful of her ancient fires,
With amorous breath his martial breast inspires;
Lifts the loved weight, tight round his helmet twines
Her loving arms, and close embraces joins.
Drawn by the mystic influence from afar,
Flies to the wedded gem the God of War.
The Magnet weds the Steel: the sacred rites
Nature attends, and th' heavenly pair unites.
 Say from what source to differing metals came
This hid affinity, this wondrous flame?
What mystic concord bends their stubborn minds?
The panting stone love's melting influence finds,
Seeks the fond metal her deep wound to heal,
Whilst love's mild pleasures tame the cruel steel."

MOLOCHITES : Μολοχὰς : *Green Jasper : Malachite.*

.

IT is difficult to discover what stone Pliny designates by
this name ; but that it was *not* our Malachite (Green Carbonate of Copper) is sufficiently evident from his words
(xxxvii. 36) : "The Molochites is not transparent, being of
a closer and coarser green than the Smaragdus (spissius
virens et crassius), obtaining its name from its mallow-leaf
colour; praised for making good impressions of the seal,
and used as an amulet for children, on account of the virtue
innate therein against all the dangers to which they are
liable.   It is produced in Arabia."   Our Malachite was his
Smaragdus Medicus, "found of greater dimensions than
any other of the sort, of a wavy pattern, representing
poppies or birds' feathers, &c., and sometimes *resembling the
Lapis-lazuli.*" (The same piece of Malachite has occasionally
one half an *unmixed blue.*)   Now, the description of the
Molochites as of a close deep-green colour, is entirely opposed to the notion of a stone where the surface is so
curiously variegated with light and dark shades, as in the
Malachite.   Besides, the latter is much too friable a substance to serve for intagli,—nay, more, to be in request for
that purpose, an inference to be drawn from the expression
"laudata."   But these characters apply admirably to a
certain pale-green Jasper, perfectly opaque, and of a dead
surface, in which admirable intagli sometimes occur, notably

the famous Cleopatra (Marlborough).* The reason why this modification of the "Jaspis" is not classed under that head, but described by Pliny as a distinct mineral, is because a certain degree of translucency was essential to the ancient idea of that species universally. Its Arabian origin also favours this attribution; at present that peculiar variety is found in India alone.

The same inference may be drawn from the strangely-confused notice in Epiphanius (under 'Sard'): "There is also another sort, the Sardonyx, which is called Molochas, and has the property of dispersing steatomatous tumours. It has the same appearance as the first (the Sard), but is somewhat greenish or yellowish (ὑποχλωρίζων)."

The cause of the ancient name for this very dissimilar stone being applied to our Malachite, lies apparently in Pliny's allusion to the general use of the former as an amulet. Now the Malachite in the Middle Ages enjoyed a mighty reputation for its virtues in this respect, probably suggested by the mysterious figures and natural hierogly-phics to be traced by a fanciful eye upon its variegated surface. De Boot says that it was even then worn in' bracelets as a preservative of the health, "being believed to possess the greatest virtues; protecting the wearer against lightning, perturbation of spirits, contagion, and witchcraft; for which latter reason it is hung round chil-dren's necks. The superstitious, and magicians who think themselves the only wise people, engrave upon it the figure of the sun, and thus believe themselves secure against .

---

* In fact there is no necessity for restricting the name to this rare species, the common dark-green Jasper of one uniform shade, perhaps approaches yet closer in tint to the marsh-mallow leaf, of a blackish, sombre, peculiar green: and its almost specific employment for making talismans and amulets has already been pointed out.

(G)   N

demons and poisonous reptiles."* Such a coincidence of supposed virtues, when the ancient material was no longer forthcoming, naturally led to the transfer of the name to the other equally virtuous substitute.

Ben Mansur, after describing his five species of the *Dehne* or Malachite, observes, " This stone is only valued in Syria and the land of the Franks." He confirms my explanation of this being Pliny's Smaragdus Medicus by the statement that if rubbed with oil it gains a greater lustre : the very thing remarked by Pliny touching the former, " those not perfectly green are improved by the application of wine and oil." To remove all doubt as to the mineral called Dehne, the Persian naturalist concludes with, " If ground down with oil and natron (soda), the purest copper is obtained therefrom (by fusion)."

Antique camei in Malachite, though extremely uncommon compared with the frequency of modern works in this soft material, nevertheless do exist. Amongst the Pulsky Gems is the most lovely bust of a Bacchante, belonging to the best period of Roman art, still retaining on portions the thin hard patina of brown oxide with which its surface was entirely incrusted when it came into the hands of the present owner—a convincing proof of the ages that must have elapsed since its first concealment in the earth.

---

* The choice of this particular sigil for this particular stone excites the suspicion of some traditional connexion in this case with the Gnostic worship of the sun-god, the pantheus *Abrasax*, whose peculiar vehicle was the green Jasper, according to me, the Molochites of those times.

MURRHINA: Moppla: *The China Agate.*

ALTHOUGH this substance is continually alluded to by the writers of Roman imperial times as the most prized ornament of the dinner-table, yet the only description they have left us of its appearance and nature is that to be found in Pliny (xxxvii. 8). " The East sends us the Murrhina. They are found in several places, all little known, and lying within the Parthian dominions, but the finest sort in Carmania.* They are supposed to be formed from a liquid hardened by subterraneous heat. In superficial extent the pieces never exceed that required for a small dish (abacus), and in thickness seldom suffice for a drinking-cup like that just mentioned (of Annius). They have a lustre without any strength, or rather a polish (nitor) than a lustre. But their value lies in the variety of their colours, the spots (or patches) suddenly turning themselves around into purple and opaque-white, and a third made up of both ; the purple, as if by a transition of hue, becoming fiery, or the milkwhite part turning red. Some principally admire the extreme parts in them (*i. e.* the edges of the vases), and a certain play of colours like that seen in the rainbow." Hence Martial makes his amateur cry for " *maculosæ* pocula murrhæ" (x. 80). " Others prefer the spots to be opaque and fatty (pingues): any transparency or paleness is considered a defect. So are marks like salt, and warts, which

* In the 'Periplus' Murrhina are also enumerated amongst the other exports from Barygaza, or Baroach, the emporium of the province now called Malwa.

do not indeed project, but are embedded in the substance, and that very often. The material is somewhat recommended also by its agreeable smell."

Such a description would appear definite enough to enable a mineralogist to identify with ease the exact substance intended, nevertheless upon no one point in the science have so many diverse theories been propounded. De Boot, after citing the then prevailing notion that the Murrhine vessels were Chinese porcelain (recently made known by the Portuguese in his own days), treats it as absurd, and explains the term as designating the vases of Sardonyx, of which he had seen antique fragments. In the last century the French archæologists, headed by Mariette, adopted the porcelain explanation, and discovered in Pliny's "purple and white spots" an exact description of the paintings ornamenting Chinaware ! It is impossible, however, to conceive anything more preposterous than the supposition that an acute observer of nature, like Pliny, and with his knowledge of art, could have mistaken the Chinese drawing, however grotesque, but evidently *laid* upon the surface by human skill, for the natural spots and veins *in* a parti-coloured stone. Old Dom Doublet could have taught Mariette better, by his apt description of a vase in the Treasury of St. Denys: " Un calice très exquis, fait d'une très belle *Agathe*, gaudronné par dehors ; admirable par la beauté et la variété des couleurs qui s'y sont trouvées naturellement esparses çà et là en façon de *papier marbré*.' A comparison affording an undesigned translation of the *maculosa murrha*, above quoted. But these antiquaries disregarded common sense, being so completely led astray by Martial's " murrhina *picta* " (a mere poetical allusion to its varied colours), and above all by Propertius with his

" Murrheaque in Parthis pocula *cocta* focis " (iv. 5).
"And Murrhine vases baked in Parthian fires."

This last description is merely the same thing, expressed in his usual farfetched way, as Pliny's, " a liquid substance solidified by subterranean heat." And in this he was wonderfully correct, for the Porcelain Jasper, a cognate material, is now pronounced to be clay metamorphosed by volcanic action. Or if we choose to take the " Parthis focis " literally, the poet may possibly allude to some method of improving the colour of the stone by heat, as still practised by the Indians with the Carnelian.* These same antiquaries must have wilfully shut their eyes to another line of Propertius (iii. 10), where he makes it synonymous with the Onyx—

" Et croceus nares Murrheus ungat onyx."

Dr. Hager considered Murrhina to be the Yu-stone or Jade vases, in all ages especially valued by the Chinese, which *might* have been brought by caravans into Carmania. Veltheim, seriously taking up a jocular remark of Lessing's, maintained that the Murrhine vases were nothing else than those so commonly seen now, made out of that dull-brown mineral streaked with white, the Chinese Steatite (or Speckstein). Böttiger does not attempt to decide the question, but rather inclines to the notion that they were of coloured glass. Corsi thought he had discovered all the required peculiarities in the Fluor Spar or Blue John of Derbyshire; but it is very doubtful if that substance was known to the Romans at all, the only mines that produce

---

* " To this day in the neighbourhood of Broach, nodules of Onyx are dug in the dry season from the beds of torrents ; they are then of a dark olive green inclining to grey ; after being exposed to the sun to dry, they are packed in earthen pots with dry goats' dung, which is set on fire. When removed, after cooling, the stones have changed in colour, often to rich hues of orange and hyacinthine red ; and the more ornamental of the mottled Onyxes that come from Cambray are those thus artificially beautified."—*Ed. Rev.* (July, 1866). See ONYX, p. 230.

the beautiful quality fit for ornamental purposes being situated in Cornwall and Derbyshire. Mongez, basing his decision upon the iridescence above noticed, proves it must have been Cæholong, or Semi-opal, but the large dimensions specified of the antique pieces entirely controvert such an explanation.*

Strange to say, the first English translator of Pliny, old Philemon Holland, has come nearer the mark than any, by rendering the word as *Cassidoine*, or Calcedony; in fact he is perfectly correct, if we take that term in its fullest sense as the generic designation of the Agate family. The only mode of arriving at the true solution of the question is by the careful examination of ancient remains, more particularly those exhumed in Rome itself. For if the whole vessels of an imperishable material were so abundant there during the four centuries of the Empire, as contemporary allusions lead us to believe, it is a logical consequence that their fragments at least must be as plentiful in the same place at the present day, since no possible circumstance could have swept them entirely out of existence. Now what is actually the case? Fragments of bowls made of Agate (but of no other irregularly-coloured stone) are turned up in abundance in the soil of the ancient capital, and often of a radius that bespeaks the extraordinary circumference of the perfect vessel. Such pieces, if not large enough to be preserved as antiques, are cut up into brooch-stones; and every spring furnishes the Roman lapidaries with an inexhaustible supply. Perfect vessels, from the fragility of the material, are rare, yet a comparatively large number are yet in existence. Of these by far the most magnificent example is to be seen amongst the Townley Pastes

---

* Besides, not one of the four stones proposed by these writers present distinct *patches* of one colour upon a ground of another, which was the grand distinction of the Murrhina.

(Brit. Mus.), a flat shallow dish with two slightly projecting ears angularly cut; fully a foot in diameter, and most exquisitely polished. The substance itself exactly corresponds to Pliny's description, being a reddish opaque purple, diversified with pure milk-white, the colours intermingling in the most intricate patterns. The stone is indeed a China Agate; for Agates present all possible varieties of colour : they occur with shades of the Sapphire passing through their white, or with well-defined bands of the brightest opaque and translucent tints; but the sort first mentioned has the best claim to be declared the true Murrhina. I have myself remarked all the changes noticed by Pliny, exhibited in a large Agate *trulla*, the colours being a partly transparent, partly opaque white, slightly opalescent, spotted with chocolate-brown, all going through the most singular variations as the light is allowed to pass through the sides at different angles. And what strongly supports this view is the fact of so many antique glass bowls and saucers being found, both whole and in fragments, which are very exact imitations of darkly-clouded Agates; for Pliny expressly mentions (xxxvi. 66) amongst the varieties of coloured glass manufactured in his day, one counterfeiting the Murrhina. A conclusive proof this that the latter substance was *not* artificial : indeed Pausanias, speaking of the water of the Styx (a most deadly and subtle poison), says (viii. 18) that it breaks to pieces not only glass vessels, but crystal, murrhina (μορρία), and all else made by man out of *stone*. Besides, owing to its common use for the most approved kind of drinking-vessel, it came to be designated absolutely as "Gemma," and the "potare gemma," so frequent in Martial, signifies this alone; "gemmati calices," on the other hand, implying gold set with precious stones. Briefly to sum up this argument : the ancient writers name Onychina, Crystallina, and Murrhina, as in equal request amongst the upper

classes of their times. Of the two first, undisputed representatives are discovered amongst their remains; of the
third none whatever, if the Agate-vases (which also occur
in the due relative proportion) be excluded.

Pliny, however, could not have been mistaken as to the
real character of the mineral; for he states that Pompey
brought it to Rome both *in the rough* and worked up into
vases (lapides * et pocula); which he dedicated to the
Capitoline Jupiter. Now Appian records that 2000 vessels
of the " gem Onychitis" formed part of the treasures of
Mithridates captured at Talaura, which shows that the
Greek *Onychitis* (not Onyx) was the same as the Roman
Murrhina, for Pliny expressly says that it was their conquest of Mithridates that first set the Romans mad after
gems and Murrhina.† Inasmuch as the district producing
the stone in its greatest perfection lay within the dominions
of this king, it was to be expected he should have engrossed
the finest specimens of the worked-up mineral. Immediately after this it came into general use in Rome for
dishes, or rather plates for food (abaci escarii), a form for
which it was best suited in consequence of the thinness of
the layers. Pieces, however, were obtained of extraordinary
superficial dimensions, for among the rarities displayed in
Pompey's triumphal procession was a draught-board, four
feet long by three wide, formed out of only two slabs

* Jan, however, reads here " capides," wine-jugs.
† These Onychitis vases were mounted in *gold ;* a sure proof they
were not of the common Alabaster originally so termed. " Murrhina "
was most likely the native word for the stone, and adopted by the
Romans for the sake of distinction. Posidonius the philosopher (quoted
Ath. xi. 89) seems to be describing the same rare collection as ὀνύχινοι
σκύφοι, of which there were found "nests," συνθέσεις, up to the size
of two cotylæ (nearly a pint). These small dimensions, quoted as exceptional for magnitude, equally prove that the precious Onyx, or
some such stone, is here intended, not the common Oriental Alabaster,
obtainable in masses of any size.

(e gemmis duabus), a magnitude never afterwards to be
obtained, or anything like it. For the size of the pieces in
Pliny's days was never beyond that required for a small
dish, and the *trulla*, especially particularised as the usual
form of vessels in this substance, exactly resembled both in
size and shape a modern breakfast-saucer. In spite of their
high price, these vessels were accumulated in prodigious
quantities by wealthy amateurs. Those belonging to a
single noble—(Annius, once Consul), notorious for having
paid 70 sestertia (70*l*.) for one cup, "which certainly,"
remarks our author, "did hold three sextarii" (pints)—
being confiscated by Nero upon the owner's death, were
sufficient when set out for public exhibition to fill a theatre
of considerable extent situated in the palace-gardens.
Amongst these Pliny *beheld* the fragments of one bowl
(scyphus) of unprecedented rarity, which, "to make the
world lament, and to cast odium upon Fortune for her
malice in breaking it," the imperial connoisseur decreed
should be preserved in a case, and thus exhibited, like the
corpse of Alexander the Great. The celebrated Petronius
could boast of a single *trulla* valued at 300 sestertia (3000*l*.),
which immediately before his suicide he smashed to pieces,
in order, as Pliny quaintly expresses it, to disinherit Nero's
table, and disappoint his expectations of so incomparable an
ornament. "But the Emperor, as became his rank, outdid
all by paying above thrice that amount * for a *capis*.† A
memorable thing that an emperor and a 'Father of his
country' should have drunk so dear!" The fragments
above mentioned as still occurring amongst antique *débris*

* Decies H. S. 10,000*l*. Our amateurs have not yet come up to this,
the 1600*l*. paid by the Marquis of Hertford for the Bernal *Rose-du-Barry*
Sèvres vase being their highest flight.

† The capis was a tall, slender jug: those in metal still preserved
being as a rule 12 inches high, it may be concluded that this crowning
glory of the Neronian beauffet attained to the same dimensions.

indicate bowls of such incredible diameter as will account for the outrageous "fancy prices" paid by enthusiastic collectors for rarities in this class.

I must venture here to differ from Winckelmann, who ('Pierres Gravées de Stosch,' p. 580) asserts his opinion that the Murrhine vases were the same as those cut out of the precious Sardonyx, in describing the portion of such a vase belonging to Card. Albani. But the two materials must not be thus confounded together, as a passage in Lampridius shows, recording as the extreme of licentious extravagance in Heliogabalus that he used both Murrhine *and* Onyx vases for the basest purposes, "in Murrhinis et Onychinis minxit :" whence it is evident that they were distinct substances. The distinction was undoubtedly this : the Onyx vases, carved in a stone of concentric layers, for the most part opaque and in strong contrast to each other, served as a medium for art, the surfaces of such vessels being always worked into cameo-designs. Such stones being usually more or less egg-shaped by nature, these vases, *alabastra, carchesia*, were deep in proportion to their width, and were infinitely more uncommon than the Murrhina, which are always mentioned as being not worth more than Crystal vessels. Thus Seneca (Ep. 123) talks of the wealthy having "mules to carry their vases of Crystal, Murrhine, and those engraved by the hands of famous artists ;" meaning by the last the ancient silver vases adorned with chasings.

The Murrhina, on the contrary, were, from the very nature of the stone, always broad and shallow, dishes or saucers, not valued as works of art, but simply on account of the material; their price augmenting with incredible rapidity as their extent surpassed a certain limit. The best sort, no doubt, was that so minutely described by Pliny, of which the noble example in the British Museum is a perfect type, but *all vases of Agate* were afterwards included under the name; all that was required being that the material should be clouded with various colours, and not present the regular stratification of the Onyx. In fact, they were exactly the same as the Agate bowls still imported from India, where the manufacture seems to be even yet going on, though in a languishing condition. At the sack of the palace at Delhi in 1858, whole chestfuls of such Agate vases, having fallen a prey to soldiers less intelligent than the captors of Talaura, are said to have been wantonly destroyed.

From the first dawn of the Renaissance these ancient gem-vases were eagerly sought after by the wealthy Italians, and imitated by their own artists, and the prices paid for them remind us faintly of the Roman extravagance in this particular. The Agate vases belonging to the French Crown were valued in the last century at half a million of francs. Their selling-price at the present day, however, is greatly fallen; they, indeed, fetch much less in London than their first cost in India.

That many of the Roman drinking-cups in coloured glass were designed as imitations of the Murrhina is manifest from their shape, that of a shallow bowl, a form adopted of necessity by the former, but by no means indispensable to the latter substance. Though in many of these bowls the glass-worker has contented himself with exactly reproducing the dark-brown, the chocolate-colour, and the white clouds of the original, yet as frequently has he aimed at surpassing Nature by mixing up the most brilliant tints his laboratory supplied in elegant waves and concentric circles.

It is very probable that such vases were imported ready made from India, as we are informed that the Crystallina were: that they were then extensively used by the Indian princes, appears from the exaggerated account Philostratus gives of the dimensions displayed by the various vessels in a precious stone, so common in that region, so rare in Greece. In the 'Periplus of the Red Sea' (written under Augustus), amongst the exports from Ozene (Ougein) are enumerated "Onyx-stones and Murrhina."

Besides their rarity, the Murrhina had another recommendation, precisely the same that made the true Chinese porcelain so indispensable when tea-drinking was first introduced into Europe,—it contained a scalding liquid without itself becoming heated, and remained cool to the lips. Now, as the Romans were extremely partial to very hot drinks, mixing their strong thick wines with boiling water and honey,* this peculiarity of the substance was invaluable. This we learn from Martial—

> " Si *calidum* potas, ardenti Murrha Falerno
> Convenit et melior fit sapor inde mero." (xiv. 13.)

Murrhina continued to be in request down to the close of the Empire, and legal writers are continually mentioning them as distinct things from vessels of glass or of the precious metals. In the Middle Ages the name came to signify a shape, not a material, in consequence, it would seem, of the most usual make in which the antique substance appeared wrought up, viz., the *trulla*. Thus a *bowl* is termed peculiarly Murreus: old inventories have often the item " Murreus de argento."

When Chinese porcelain began to find its way into Europe,† the learned, not knowing how it was made, seem

---

* Hence the usual term for a tavern, *Thermopolium.*

† It was brought to Europe by the Spanish traders, who in the back voyage from Peru doubled the Horn, and touched at the Phillippines and Manilla. Of this an Elizabethan poet, T. Weekes (1597), avails himself to produce this perfect specimen of the bathos,

> " The Andalusian merchant that returns
> Laden with cutchinele and china-dishes
> Reports in Spain how strangely Fogo burnes
> Amidst an ocean full of flying-fishes."

to have agreed to call it Murrhina, upon the axiom that things equally unknown must be identical. Long before that time the Italians had their *Porcellana* (so called from the pounded shell *porcella* entering into its composition), a mere glazed earthenware, the secret of which had been taught them by the Moors: which was the reason that afterwards "Porcelain" was improperly applied to the Chinese manufacture, though of a totally different nature. Garcias ab Horto affords a good illustration : "They find a green kind of Jasper, out of which Murrhina are made (called Porcellana) of so bright a green that they might be thought cut out of the Emerald : and such is probably the material of the Sacro Catino at Genoa. A Murrhine vessel of this kind was once offered to me for 200 pardoãs (rupees) ; had it been a true Emerald I could hardly have got the thousandth part of it for that price."

Lambeccius figures ('Bib. Cæs.,' vol. i.) of the actual size a very elegant patera with twisted handles, designed in a pure classical style, and then (1665) believed to be hollowed out of one entire Emerald, although more than a foot in diameter, measured over the handles. This therefore must have been, like the Sacro Catino, formed in a brilliant green paste, for certainly no mere green Jasper could in that age have been mistaken for the true Emerald, especially in Vienna, the school of the most learned amongst the early mineralogists. The Imperial Library *then* boasted of the most magnificent example of the true Murrhina ever produced by ancient skill, or preserved, as a most sacred treasure, protected also by superstition, through all the intermediate ages between the Roman and the Hapsburg Cæsars. Lambeccius calls the material Oriental Agate, which exactly agrees with my foregoing definition of the real signification of the name Murrhina. The form was a patera with gracefully twisted handles, and much resembling the

Emerald vase in general design; but the size was perfectly
incredible, the diameter over the handles measuring a
Viennese ell all but two digits (about 27 inches). This
also, to remove all doubt, is given of the full size upon the
same gigantic folding plate. But what then gave it supreme
importance, and probably had conduced chiefly to its preservation through the wreck of ages, was the inscription,
"not made by hands," but traced by the finger of Nature
(as the imperial librarian states, and evidently in good
faith),* in the colours of the stone, B. XRISTO. RS. XXX.,
interpreted as "Christus Beator Orbis." To produce the
effect of natural shades in the substance these letters must
have been written with nitric acid, which leaves a slight
but indelible opacity where applied, of which there is an
example (Marlborough) known to myself, the name and
titles of a former owner, ANDREAS CARRAIA, traced round
the edge of a singularly fine Cameo (Constantine with
helmed head). The mixture of Greek and Roman capitals
betrays a Byzantine date, and seemingly refers to the con-
secration of this unparalleled relic of imperial splendour to
the service of the Altar. No tradition is recorded of its
origin: but as it had been "for many centuries" in the
same possession, everything leads to the belief that it found
its way to Vienna after the Frankish conquest of Constantinople, when so many grand Camei became dispersed
throughout Europe. The existence of such a natural prodigy as a slab of Agate of the dimensions this patera demanded renders the vast prices given for similar rarities
under the Cæsars no longer a thing to be wondered at. The
shape of this dish is precisely that of the trulla, and of the
class of vessels more peculiarly formed out of the Murrhine-
stone, as its natural capabilities dictated. The Emperor

* Adding, "Though man be silent, the very *stones* proclaim the
glory of God!"

Leopold exhibited this Agate to Lambeccius with his own hands, as the most important of all his artistic treasures; but to him also the miraculous inscription constituted its principal value.

The front of the altar in the Chiesa del Jesù, Rome, is formed of a single slab of what the archæologists there pronounce the long-sought Murrhinum, found in the ruins of an ancient house in the Forum. ' "It answers the description in Pliny very well. It is purple in colour, with strata of a dull white running through it : on the edges of the white there is a slight iridescence. It bears a resemblance to fluor-spar, but it is *not* fluor, for, as Sibilio at Rome showed me, fluor-spar is easily scratched, whereas the so-called *Murra* cannot be scratched, more particularly in the *white.* This white vein is a pure horn-stone."— (H. M. Westropp.) Brückmann possessed a cameo in a similar composite, one stratum silicious, the other calcareous. Such a mineral was doubtless classed amongst the Murrhina by the Romans, but, as certainly, did not exclusively engross that appellation.

NAXIUM : Σμύρις : Σμίρις : *Emery.*

ALTHOUGH the use of this mineral is coeval with the first invention of gem-engraving, which indeed in its proper sense cannot be effected without its aid, yet the earliest extant mention of it by a specific name is that made by Dioscorides (a contemporary of the Triumvirs), who has (v. 165), " The Smyris is a stone with which gem-engravers polish the gems." Theophrastus (44) mentions the same use of the mineral, but appears not to have had any distinctive denomination for it, his words merely being, "And again, the stone wherewith they engrave signets is the same as that of which whetstones are made, or else similar to it, and the best kind is brought from Armenia."

The very chapter in which Theophrastus had given some details of the process is unfortunately one of the most corrupt and mutilated in the whole treatise ; but it may be gathered from what remains that he expresses his surprise at the fact that the substance in question can be cut up and shaped by a steel tool, and yet can bite upon gems that the steel will not touch.

Ἀκόναι with the Greeks, as *Cotes* with the Latins, were terms used indifferently to denote the agent with which steel tools were sharpened and " *hard stones* "* engraved, inasmuch as a piece of compact Emery was employed in both

---

* ' Hard stones ' is the technical term in the lapidary's art for all the species which resist cutting by a steel tool, however hard its temper.

cases and in much the same manner, that is to say, as a file, not as now in powder applied by another mechanical contrivance. Dioscorides prescribes (v. 167) "the dust rubbed off a Naxian whetstone by the steel sharpened upon it." Pliny, after stating (xxxvi. 10), "For polishing marble statues, and also for engraving and *filing down* gems, the Naxium long held the first rank; thus are called the whetstones (cotes) produced in the isle of Cyprus," and (xxxvii. 32) "the Peridot is the only precious stone that yields to the file, all the others are polished by means of Naxium and whetstones." After this, when he comes to treat (xxxvi. 47) of the whetstones used for tools, he repeats the same observations as to the Naxian, and the superiority of the Armenian. In another place (xviii. 67, § 5) he mentions the ancient Italian mowers as knowing no other whetstones than those brought from Crete, and other places beyond the sea. For these oil was necessary, and therefore the mower moved along with a horn full of oil tied to his leg. But Italy afterwards supplied whetstones which required water alone, and acted like a file upon the scythe.

For marble-sawing the best sand was imported from Ethiopia, the next quality from India, "peteretur etiam in Indos quo margaritas quoque peti severis moribus indignum erat" (xxxvi. 9). This so-called *sand* must have been Emery-powder; and in the fact of its being gotten in Ethiopia lies a very good reason for supposing its use known to the old Egyptians, and confirms Sir G. Wilkinson's explanation of the manner in which their primitive sculptors contrived to master the most obdurate materials. That country, in truth, seems to have given the name to the mineral, its export, for *Zmiri* in Ethiopia is often referred to as a gem-producing locality.

Those primeval monuments of gem-engraving, the Assyrian cylinders in Loadstone, sufficiently explain Pliny's

(G)
o

expression of *filing* a gem, the designs having evidently been cut into the circumference by rubbing with the edge of a fragment of Emery. It is at a more advanced period of Assyrian art that we perceive the neatly turned and regular indentations marking the application of the drill. The backs also of most really antique intagli of all periods show by the deep furrows upon them, but imperfectly concealed by the lustrous polish subsequently imparted, how the gems had literally been *filed* into shape by the rubbing with an Emery-stone.

Hence we discover why Armenia should be famous for the production of the best material for this purpose; it was the source from which the inventors of the art drew their supplies of the indispensable agent. As the knowledge of gem-engraving spread from Assyria towards the coast of Asia Minor, the artists carried with them a supply of the Armenian mineral. Theophrastus shows that in his days it was still imported from that region into Greece, although the isle of Naxos then as now possessed inexhaustible mines. From the preference given by the Greeks, at the time when the Glyptic art had reached its highest point amongst them, to the Armenian kind, there is reason to suspect that the latter was a purer Corundum than the Naxian, and therefore more efficient in engraving. Modern times supply an analogous case in the same process. The Adamantine Spar, when introduced shortly before 1790 (Raspe) into the European lapidary's atelier from India, where it had been known from time immemorial, was eagerly received and immediately adopted universally as an enormous improvement upon the old-fashioned Emery-powder. It is self-evident that the Greeks would not have taken the trouble to import from a long distance, Armenia, precisely the same mineral as they were at the very time obtaining from their immediate neighbourhood, Naxos, with the smallest expenditure of labour, and in any quantity required.

The process known to the most barbarous tribes of primeval times, that of drilling holes through stones by the long-continued turning of a stick, constantly supplied with sharp sand and water, upon the same point, has left us many memorials in the Celtic axe and hammer-heads, where the shape of the widely-splayed helve-holes plainly indicates the agents brought into play. These are abundant in regions which do not produce Emery; but sharp sand, aided by the unlimited expenditure of time, effects the same result.*

The Assyrian gem-engraver soon perceived the advantage to be derived from an instrument that could hollow out his figures in hemispherical indentations, exactly formed to the size required, in addition to the simple straight cuts which alone he could produce by the original method of filing with the fractured Emery. The perforations traversing the length of all these cylinders prove that already, as soon as they were invented as signets, the principle of the drill was well known. These holes are bored so truly that we discover not merely that they were done with a metal wire, instead of a stick, charged with Emery and oil, but that this wire must have been turned by means of a bow, in order to work thus evenly, and not to run the risk of splitting the cylinder. It could not have been long before

* The Mexicans had no iron, the New Zealanders no metal whatever, yet the former cut the Amazon-stone, the latter the Jade, both amongst the hardest stones known to the lapidary, into the most elaborate and highly-finished monsters. La Chaux mentions a very instructive case of a return to a soft medium for the application of the cutting particles amidst the most elaborate appliances of the art : Guay gave the finishing touches to his *chef-d'œuvre* the bust of Louis XV. in cameo, by working the diamond-powder with the point of a *quill* into lines otherwise inaccessible. Intagli also are polished internally by means of a *leaden* point similarly charged.

the engravers ascertained that the same principle applied by means of blunt points varying in thickness would enable them to sink almost the whole of their design with extreme accuracy, only requiring a slight finish from the cutting medium. This is why many of the archaic Babylonian class exhibit in their mode of execution as extensive, or rather exclusive, employment of the drill, as marks the largest proportion of the Etruscan scarabæi, which still retain the stamp of their Asiatic origin in their *technique*, though not in their subjects.

It must be here explained, that the metal point takes no part in cutting the intaglio, but merely serves as a medium in which the acuter particles of the pounded Corundum embed themselves, and bring an infinite number of cutting surfaces, continually renewed, to bear upon the gem. The softer the metal the larger proportion of these cutting particles does it take up and carry; hence a bronze drill will work faster than one of steel. The modern gem-engraver sets all his drills and cutting tools, however differently shaped, to work, by making them revolve horizontally by means of a lathe driven by the foot, whilst he presses the gem cemented into a handle against the cutting edge. These cutters are, as they are wanted, fixed in a spindle driven by the *wheel*, which last gives the name to the entire contrivance. On the contrary, the ancient engraver, reversing the operation, brought his drills to bear upon a fixed surface, and turned them merely by means of the bow moved by his right hand, whilst he directed the drill with the left. Thus only is it possible to understand how the ancients were enabled to operate upon those enormous camei, the extent of which presents to the practical eye difficulties almost insuperable by the modern process. More than one artist could thus have laboured simultaneously upon the same slab of Sardonyx, the simple drill and bow requiring so little room for their application. Indeed every really antique cameo on a large scale retains traces that show distinctly how the superfluous parts of the upper stratum have been excised by means of circular perforations.

Sir G. Wilkinson is also of opinion that the Egyptians performed their stupendous sculptures in Basalt, Granite, and other hard rocks, that now speedily turn the edge of the best steel chisels, by using bronze tools constantly sup- ⟩ plied with Emery-dust, the tool serving only as a medium for driving the cutting particles into the stone. The interior of the hieroglyphics certainly has every appearance of having been *bruised* in by some such agency. And connected with this application of the mineral, a circumstance arrests our attention which at the first view must strike every one as utterly at variance with the usual order of things. It is the phenomenon that so many, and widely separated, races should at the first dawn of their civilization have attained to the fullest proficiency in all the mechanical processes of what is assuredly one of the most refined of the arts belonging to luxurious opulence, the art of engraving in " Hard Stones." The strongest illustrations of this fact, and those too the furthest removed from each other in time as well as in place, are to be discovered amongst the remains of the Assyrians and of the Aztecs: the cylinders of the first in Hematite and Agate, the idols of the latter in Jade and Amazon-stone.

But the apparent inconsistency vanishes upon a nearer insight into the " prehistoric annals " of the progress of the human race in the arts of life. The steps, at least in its earliest stages, have been the same pretty nearly for all nations as well as for all ages: for one common instinct has suggested the appliance of similar natural materials to the supplying of similar common wants. The weapons, tools, jewels, of primitive man have ever been fashioned out of stones: and the stones again have been fashioned to their object by means of others of harder texture. The recent discoveries amongst the Lake-dwelling of the original Celtic population inhabiting Switzerland have supplied abundant illustrations of the method (heretofore a mystery to all who considered the matter) by which the axes and chisels of Serpentine and Schist were first shaped out of the block by sawing with a sharp edge of flint; a mineral, it must be remarked *en passant*, not indigenous, but imported from Gaul for the purpose. The hammer-heads of the same locality also furnish examples of the operation of the primitive drill in their neatly-bored helve-holes, supposed to have been worked out by the continued rotation of a cylindrical flint assisted by sharp sand. The unfinished specimens disclose the curious contrivance by which this borer, whatever its nature, was made hollow so as to leave a core for itself in the block to which it was applied, and effectually to obviate its shifting its position.

Now, the historians of the conquest of Mexico particularise, as a thing very remarkable, that the natives, though at that time well acquainted with the use of cutting-instruments made of hard-tempered bronze, employed nothing except flakes of Obsidian for sculpturing the reliefs in Alabaster and other calcareous stones, in such great request for the decoration of their buildings. Such cutting-agents, like the flint-saws of the primitive Swiss, could only produce their effect by means of continuous rubbing, not by impact; in other words, they acted precisely in the same way in which the "Armenian stone" worked out the minute designs upon the harder gems to which it was applied by the Babylonian seal-engravers. From carving a hard coarse stone to carving one that was fine and precious, but not more refractory to incision, was a very natural step. Ornaments whose chief importance, it may well be conjectured, lay in their virtues as talismans, would necessarily be formed of the most costly materials at the command of the individual intending to decorate and secure himself with the same appendages. The Mexicans were found by the invaders cutting the brittle Emerald and Amazon-stone, and the much more obdurate Jade, into curious and fantastic shapes with a skill that excited the admiration even of the accomplished Italian jewellers of that epoch. In this operation it is stated that bronze tools were used, but the true agent was the " silicious dust " also mentioned, unintelligently, by the same old Spanish historian. For this delicate style of work metal-points were required to enable the engraver to apply the cutting particles with nice precision to every portion of his work: in other words, the Aztec gem-engraver, taught by long experience, had furnished himself with exactly the same tools which had served the Etruscan and Archaic Greek artist in executing their finished works some two thousand years before. And the " silicious dust " must have been a powder either provided by nature, or artificially prepared from some form of the Corundum-stone; for no mere *silex* in powder would be capable of acting upon such a material as Jade. With the same means the Mexican lapidaries are described as polishing and shaping the hard Obsidian into mirrors which answered their purpose with tolerable accuracy, and also as carving Porphyry, Basalt, and Granite into elaborate figures. As a proof of the effective character of the instruments, as well as of the dexterity with which they were applied, lions in granite are known of Mexican workmanship holding in their mouths moveable rings carved out of the same block as the animals themselves. Of the most celebrated specimens of their proficiency in this art a full description will be found in ' Precious Stones,' p. 299.

The circumstance distinctly specified that the carvings in the harder rocks were done with the same instruments as those used in working the precious stones, affords a singular confirmation to the theory already quoted, that the Egyptians effected their stupendous works in similar materials by the application of the same means. And again, to complete the analogy between the processes of the two nations, the use of Obsidian flakes in carving the softer stones is the counterpart of the Egyptian use of flint in incising their steaschist scarabei : an instrument not merely to be conjectured from its traces left in its productions, but indicated by the express declaration of Herodotus, that the Ethiopians in the army of Xerxes had their arrows pointed with the same stone that they used for engraving signets. And what this stone was is placed out of doubt by the discovery of flint-headed arrows in Egyptian tombs.

There is a Rabbinical tradition testifying to the extreme antiquity of the employment of the *Smyris* in gem-engraving. Moses, they tell, engraved the stones of the Rationale by means of the blood of the worm *Samir*, so powerful a solvent as to subdue the hardest gem, and leave a hollow wherever the characters had been traced therewith. In after ages, when Solomon was about to build the Temple with stones untouched by the tool, the source whence Moses had obtained the Samir had been lost in the darkness of antiquity. But Solomon's wisdom was equal to the emergency, and speedily suggested a method for its recovery. He inclosed an ostrich-chick in a glass vase and set a watch upon it; the parent-bird, finding it impossible to break the vase by force, flew off to the desert, and returned with a supply of the wondrous reptile, the application of which speedily dissolved the glass and released her young one. By repeating the stratagem, a sufficient supply of the potent menstruum was obtained. The root of "Samir" is evidently the same as that of "Smyris;" and there is another point of analogy between the significations. *Samir* is explained by some Rabbins as meaning the Diamond or *Adamas;* now it is almost demonstrable that the Adamas of the early Greeks, before the Indian Diamond had found its way among them, denoted every kind of Corundum, whether precious stone or instrument of art. Again, *Samir* is the word Ezekiel uses (iii. 9) as the most expressive synonym for hardness: " As an *Adamant* harder than flint have I made thy forehead." But when the precious stone is intended (in the Rationale) a different word, *Jahalom,* is employed. The Jews, whose artists were all called in when required from foreign nations, Phœnicia chiefly, had preserved a tradition of the use of *Samir* in the cutting of their famous Breastplate, and, with their usual love of the marvéllous and the absurd, invented the rest of the legend. There seems, however, always to have prevailed a notion amongst Orientals that gems could be softened by chemical means: a trace of this appears in Pliny's repetition of the fable that the Diamond could only be broken after maceration in goat's blood. Certainly to persons unacquainted with the art of engraving, the ability to sink designs deep and rounded in outline into the most obdurate of natural productions, gems, must have appeared miraculous, and only to be explained by the existence of some secret means for softening the hardness into a plastic consistency. Such a wonderful property would be naturally ascribed by an Oriental to the blood of a serpent or worm, creatures figuring so largely in magical operations. From such a prepossession comes it that we find the Indian Diamond first made known to the Greeks by Sotacus as the *Dracontias,* and only found in the serpent's head.

Earth-worms indeed are recommended as a menstruum for softening glass by Heraclius, who wrote a treatise, ' De

Artibus Romanorum,' probably in the seventh century, in a barbarous and often unintelligible jargon.   However, if we choose to believe his assertion,

"Nil tibi scribo quidem quod non prius esse probassum,"

the following recipe would decide the question :—" Collect fat earth-worms as turned up by the plough, vinegar, and the hot blood out of a big he-goat fed on strengthening herbs ; mix all together, and so anoint the bright shining bowl, and then engrave upon it with fragments of the hard stone called Pyrites." This was the method, as he prefaces, directed by the art.   (See VITRUM.)

*

LAPIS NEPHRITICUS: *Jade.*

THIS singular mineral is a combination of magnesia and silica, with small proportions of alumina and the oxides of iron and chrome. In colour it varies from a soapy greenish white, with a waxy surface, to a clear agreeable olive—the most esteemed shade. The Egyptian kind, Corsi states, is the greenest of all, approximating in beauty to the Chrysoprase. This substance is excessively hard, tough, and difficult to work, almost insuperable by emery, and requiring the employment of diamond powder in the operation. It therefore appears to have baffled the skill not merely of the ancients, but even of the difficulty-courting artists of the Revival, no work exhibiting the well-known style of that period existing in Jade; and yet the latter had every inducement to essay this material in the high reputation it enjoyed in their own times.

This reputation rested upon its well-accredited virtue as a specific remedy, or rather as a prophylactic, against all diseases of the kidneys, which gives it its present popular name, corrupted by the French from *pietra di hijada* (kidney-stone) as the Spaniards had christened it. They had introduced the stone together with (as it would appear) the belief in its peculiar virtue from the New World very soon after its discovery. In its efficacy even the practical De Boot was evidently a firm believer; and the prices quoted by him equally testify to the general faith in its medicinal virtue; a piece no larger than a half-thaler selling at 100 crowns, whilst a mass of sufficient dimensions to make a good-sized cup, then in the hands of the imperial jeweller, was valued at 1600 thalers.*

The same author states that it was then brought from New Spain, and quotes Monardes (a Spanish physician, whose treatise was published in 1567) as to the ancient estimation in which the natives had held it: "The Indians wear the Nephrite, shaped into various figures, some of fishes, some like birds' heads or parrots' beaks, others spherical like rosary-beads, but all of them perforated, because from time immemorial they wear them as pendants, against all diseases of the kidneys and pains in the stomach, in which complaints these ornaments are highly esteemed for their efficacy." De Boot again mentions (under *Smaragdoprasius*) that the West Indians, the Caribs, wrought that particular stone, which some held to be the only genuine Nephrite, into a cylinder, enlarged at one end, of the size of the little finger, which was worn by their chiefs as a badge of rank, through a hole pierced in the lower lip. He also asserts a very singular thing, that the true Nephrite was then found in Spain and also in Bohemia; a fact which he adds was but little known, because the ignorant lapidaries mistook it for the Smaragdoprase, or for the green Jasper. Strange to say this most obdurate material has held the same place amongst certain barbarian races, though the most remotely situated from each other, that was occupied by the Onyx amongst the most refined nations of antiquity. The Mexicans, when first discovered, employed the Jade as their chief jewel, before the Emerald, and had

---

* De Laet mentions having seen in England a mass as big as a man's head, of a milky greenish colour; lately brought from the river Amazon, where this stone was found in the greatest abundance. The owner asked 50*l.* sterling for it: a vast sum in James I.'s days.

learnt how, by some process now unknown, to cut it with singular dexterity and neatness into the various fanciful patterns described by the old Spanish physician. It was the " Chalchivite," the "green stone," in higher estimation than any other amongst the Aztecs, doubtless on account of its superiority in hardness to the Emerald, and which, associated with the latter, now so precious, stone, and with pearls, thickly studded the mantle of Montezuma in his first interview with Cortez. Out of this same gem was carved the elaborate clasp fastening the monarch's imperial robe; for green was the colour appropriated to royalty in ancient Mexico. On account of this its supereminent value the old chronicler of the Conquest, Diaz, contents himself, according to his own account, with four pieces of this gem alone out of all the accumulated treasures of Montezuma in their division amongst the soldiers on the *noche triste*, "disastrous night" (July, 1520), when Cortez was forced to evacuate Mexico. Those who had burdened themselves with the golden ingots perished in their attempts to ford the breaches cut by the enemy in the long causeway leading across the lake to terra firma, whilst Diaz then and afterwards had full reason to congratulate himself upon his prudent choice.

Many ages later a far ruder race, the New Zealanders, were found making the same stone the badge of sovereignty, worn by their chiefs, thrust through the lobe of the ear either in the form of a hook, or of a thick cylinder, worked perfectly even, and highly polished. Like the Mexicans, they selected for making such jewels the pure translucent green species. The coarser sort was formed with equal ingenuity and an incalculable expenditure of labour into the blades of their adzes. And here it must be remarked that the most curious circumstance connected with the history of this enigmatical mineral is the established fact that it has been used for the latter purpose from the primitive ages of the world, jade axe-heads having been found along with the other stone implements of the lake-dwelling aborigines of Switzerland. The inexplicable mystery remains, whence they obtained the stone; for in spite of De Boot's statement, above quoted, modern mineralogists deny its existence in any part of Europe.

With the Chinese, however, it is the material, above all others, appropriated to the Glyptic art, such as theirs is. The clear green sort, indeed, they value above every precious stone, and employ only in the form of beads. The Emperor's own necklace, captured in the late sacking of the " Summer Palace," was composed of such spherules of the size of large cherries, placed singly at intervals marking their rarity, between several others of the finest coral, the pendant from the centre being a monster spinel, or perhaps a red tourmaline. The grayer, more waxy species, obtainable also in larger masses, serves for making idols, vases, cups, and for the broad circular plaques used for clasps to the belt, most elaborately carved into open work of foliage and flowers, in the same style of ornamentation and with the same freedom as their works in ivory. How they have obtained such mastery over so difficult a subject-matter remains a mystery to Europeans. Some pretend, but it would seem merely on conjecture, that the stone is softer when first extracted from the bed, and acquires its hardness artificially, through a baking after it has received its carving and polish. There is a story current, quite *à la Chinoise*, relating to this operation, that whenever a piece of unusual magnitude and capabilities has been discovered, the Emperor summons a chapter of artists to deliberate upon the best shape into which it can be carved. The candidate whose model is approved by the president obtains the commission, but on the terms that his work, when finished, must be submitted to a similar jury of professionals, and if not judged satisfactory, the carver has to lose his head. But inasmuch as *twenty* years are required for the completion of such a task, during all which the engraver receives a handsome salary from his employer, the very remote contingency of failure and punishment tends little to damp the ambition of the competitors for the job.*

No satisfactory evidence has come to my knowledge as to this mineral having ever found its way to ancient Rome, as it possibly might have done along with the other productions of the remotest East. No pieces of Jade ever come to light stamped with the unmistakeable impress of classic art, although I have seen a few intagli in a waxy calcedony † that might easily have been mistaken for it. Pliny indeed mentions the *Adadu-nephros,* " kidney of Adad," the Syrian Belus; but as he also specifies the " eye " and the " finger " of that deity, in the same way as he does the " Idæi Dactyli," *Jove's fingers,* it is evident that such names denote only the natural configuration of certain fossils, not their medicinal relation to the same members of the human body. Similar parts of other deities were espied by the fanciful Magi in the accidental shapes or markings of different stones. Most curious of all was the *Hermuaidoion,* so called, " ex argumento nominis," from its representing the organ, that deity's

* In Jade we have the most remarkable monument, taken all in all, that the Glyptic art of any age or nation has ever produced. This is the immense Tortoise found in the bank of the Jumna, and now displayed in its appropriate place in the mineralogical gallery of the British Museum. It is carved with exact fidelity to nature, and perfectly polished, out of an unparalleled block of fine olive-green, agreeably clouded with lighter shades.

† Mr. H. M. Westropp informs me that he possesses a bit of Jade picked up in an excavation on the Palatine; and that the stone goes in Rome by the name of *verde di Tarquinia.* But this very appellation, indicating it to be a *native* of that locality, throws great doubt upon its identity with the Oriental species.

especial symbol, depicted by nature upon either a black or a white ground, and inclosed within a golden circle. The idea reminds us of its counterpart, the *Hysteropetræ*, appropriated to the other sex, that excited so much wonder in the mediæval geologists, like Agricola, who describes them as "lapides nigricantes et duros qui pudendum muliebre exprimerent." They were found about Marienburgh and Ehrenbreitstein, and were regarded, Uffenbach tells us, as miraculous memorials of the unsullied virginity of St. Ursula and her eleven thousand maiden companions. The specific remedial virtue of the Jade is attributed by Orpheus to the *Crystal*, if worn tied about the loins of the person suffering from nephritic complaints.

OBSIDIANUM: *Opsianum:* Λίθος Οψιανὸς.

THIS substance, a vitreous lava, whence its old name,
" lava-glass," was called after a certain Obsidius or Opsius,
who first discovered it in Ethiopia (xxxvi. 67). " It is
extremely black and sometimes translucent (in thin pieces),
and, inlaid in chamber-walls in the form of mirrors, it
reflects shadows instead of images." By a strange coincidence, the Peruvians, when first discovered, had Obsidian
(Itzli) mirrors in general use. One of these Obsidian
mirrors seems to have found its way to England, and to
have been converted into the identical " black stone into
which Dr. Dee did call his spirits :" for so it is described
in the Peterborough Catalogue. It had been long, probably since the conjuror's death, in the possession of the
Mordaunts; after other changes of owners, in Horace Walpole's; and now is in the Londesborough Collection. The
strange shape to which the stone is cut betokens an Aztec
origin. This may dispute with the crystal sphere, already
noticed, the honour of being the instrument used by Kelly,
Dr. Dee's coadjutor; whereof Butler :—

> " Kelly did all his feats upon
> The Devil's looking-glass, a stone ;
> Where, playing with him at bopeep,
> He solved all problems ne'er so deep."

" Many persons (adds Pliny) make signet-stones out of it."
This fashion does not seem to have lasted long, for antique
intagli in Obsidian are extremely rare. There was, how-

(G) P

ever, in the Praun Collection a Cock-Chimæra, on a large
Obsidian, with a Gnostic design on the reverse. The Black
Agate, much employed by the Roman engravers of his and
the following century for the representation of masks, may
be distinguished from this stone by the character that,
however dark its colour, it always exhibits a reddish
translucency when held up against a strong light; the
Obsidian, on the contrary, where translucent, is colourless.

Pliny had seen figures in full relief (of Augustus) sculptured in it, the substance attaining sufficient dimensions to
allow of such employment;* and the same emperor had
dedicated in the Temple of Concord, as being something
wonderful, four elephants cut out of Obsidian; their size,
however, is not stated. Tiberius " restored to the worship
of the people of Heliopolis " a statue of Menelaus, which
had been found amongst the property of a certain former
governor of Egypt (who appears to have appropriated it
in the Verrine fashion), which Pliny quotes as a proof that
the stone had been used in sculpture long before its pretended discovery by Obsidius. According to Xenocrates
it was found not merely in Ethiopia, but in India, in Samnium, and in Spain, upon the Atlantic coast: and in this
the Greek was doubtless correct, for it is commonly found
in the vicinity of all volcanoes, the chief deposits known

* "Capaci materia hujus crassitudinis." Pliny seems, however, to
distinguish the stone from the gem in a subsequent passage (xxxvii.
65) : " Of the stone Opsianus we have already treated in the last book :
gems also are found of the same name and colour, not only in India and
Ethiopia, but in Samnium also, and as some think in Spain, on the
beach of the ocean there."

It is almost evident he has here in view the fine black opaque ring-
stone, now known as the Black Jasper, as well as the translucent Black
Agate: all the three being identical in appearance except when viewed
by transmitted light: the frequency of antique intagli in the two latter
species, and their rarity in true Obsidian almost demonstrate this. His
statues were probably of that rarest of *marbles*, the " Nero Antico."

at present being in the Lipari Isles and under Mount Hecla.

In Pliny's time it was largely imitated in glass, employed as a material for plates and dishes. A similar black glass was a favourite medium in the last century for making pastes after antique gems, which renders it often difficult to distinguish the antique engravings in the real Obsidian from these modern fabrications, unless by the test of the file. For Obsidian, notwithstanding its vitreous nature, possesses the hardness of the Garnet, and its splinters will deeply scratch into the surface of any paste, for which property Pliny recommends them as the surest test for the detection of the latter mode of imposition.

There was a curious pedantic error prevalent in the same century of calling all antique pastes *Obsidian*, merely because Pliny had mentioned the imitation of this particular stone in glass. But Pastes, generically, are termed by him "*vitreæ gemmæ*, used for setting in the rings of the populace" (xxxv. 30).

Orpheus (282) prescribes "the stern Opsian stone" as an ingredient in a curious magical recipe. It was to be mixed with amber, "the tears of the pine," myrrh, and the flaky talc, when the compound would return oracles "concerning all things about to happen, whether good or evil, and give to know whatever thou mayest desire."

The Mexicans are described by Ximenes, an early Spanish missionary, as making razors and knives out of Obsidian with great dexterity. Holding the block between their two feet, they, with a *wooden* instrument shaped like a musket-butt, struck off flakes. The modern Indians still manufacture Obsidian arrow-heads by splintering off thin flakes from the mass by means of a peculiar and dexterous pressure with the point of a goat's horn, in a manner that can only be understood by actual inspection. The Celtic

P 2

arrow-heads in flint so truly made, so exactly balanced, must have been produced by some similar application of a softer medium : it is found impossible to reproduce their make with any degree of perfection by modern tools. These Obsidian laminæ, adds Ximenes, being mounted as an edge to a wooden sword, furnished a most terrible weapon, capable of cutting a man in twain, nay more, of slaying a horse, by a single blow, after which, however, it became useless.

Another volcanic product, in great request with the later Egyptian engravers, was a *Trachyte*, exactly resembling Black Jasper, with which it is always now confounded, though distinguishable by its levity and bubbly texture. The exclusive employment of this Trachyte for talismanic subjects inclines me to recognise in it Pliny's *Antipathes*, " black and not translucent,—recommended by the Magi as a protection against witchcraft."

ONYX: 'Ονύχιον: 'Ονυχίτης: *Nicolo.*

THE name of Onyx was given by the Romans to two totally
distinct substances: a species of marble, and a silicious
gem. Pliny states this expressly: "hoc aliubi lapidis, hic
gemmæ vocabulum." As it would appear, from a circumstance hereafter to be noticed, that the marble was the
first of the two to be known under that name to the Romans,
it is properly the first to be here considered. It was the
carbonate of lime, now called Oriental alabaster, and received its original appellation from the fancied resemblance
of its clearly defined white and yellow veins to the shades
in the human finger-nail (ὄνυξ). The Greeks, as was their
wont, discovered this familiar word in the Semitic "Oneg,"
"a delight," or "the jewel" above all others; seeing the
paramount value that race have ever attached to the gem,
of which more shall be said anon. "Oneg" in the sense
of "jewel" is exactly analogous to our derivation of the
latter word from "joyau" and "gioiello."

In the republican times of Rome the *marble* was a material
of incredible value and rarity, supposed to be peculiar to
Arabia, and solely employed for making drinking-cups and
the feet of couches. Some singular illustrations of its
original scarcity in that city, and of the rapidity with
which it became plentiful there under the stimulating
influence of the fast-growing luxury of the early empire,
will be found collected under *Alabastrites.* Being con-

sidered as especially adapted for the preservation of the
unguenta, or perfumed oils, then so much in use, it was
worked up into the little vases called alabastra (or "without handles"), and thereby becoming known as the *alabastrites*, or "unguent-jar stone," lost its original designation
of Onyx, which thenceforward came to be confined to the
gem now to be discussed.

The first mention of the Onyx as a precious stone occurs
in the inscriptions of the Parthenon, dating from the Pelo-
ponnesian War (B.C. 431–404), where amongst the offerings is registered "a *large* Onyx engraved with an ante-
lope; weight 33 drachms." Theophrastus ('On Stones,'
31) uses the diminutive ὀνύχιον, indicative of its value, and
also perhaps to distinguish it from the marble Onyx; and
describes it as made up of white and dark-brown (φαῖος) in
alternate layers. This ὀνύχιον of Theophrastus and the
early Greeks was not the stone known to the Romans and
to us as the Onyx, or *Nicolo* (the Italian corruption of
Onyculus), but the gem upon which the best archaic and
Etruscan intagli for the most part occur, and which is now
commonly called the Tricoloured Agate. This appears
from the description above quoted from Theophrastus. His
definition of one of its shades as φαῖος gives the very epithet
Homer frequently applies to spring-water, and therefore
can only signify something blackish and at the same time
translucent, the actual appearance presented by water deep
and clear. Totally inapplicable are these terms to the
Onyx of the Romans, our Nicolo, where the layers are
opaque, and usually of vivid colours, blue horizontally
superimposed upon black; but they exactly describe the
appearance of what is universally called by dactylio-
graphers the "Tricoloured, or Banded, Agate," that so
favourite material with the early Glyptic art, and which the
primitive lapidaries invariably cut *across* the strata, so as to

obtain two bands of dark brown, one lighter than the other, and separated by a middle zone purely colourless and transparent. This stone, it is evident, enjoyed the highest reputation for the purpose of signets, until shortly before the rise of the Roman Empire, when the brilliantly-coloured Indian Sard came into competition with and completely banished it from the fingers of the fashionable. As a proof of this, it may be safely asserted that not a single imperial portrait is known to exist in such a material. On the other hand, equally into request with the Romans as the all-prevailing Sard came the yet more recent Nicolo, a gem completely unknown to the Greek engravers, as Caylus has long ago observed, and the experience of all gem collectors will confirm his observation.

Isidorus of Seville has noted that the old Latin name for a signet was *ungulus;* for which he offers the far-fetched explanation that it was so called " because the gem covers the ring in the same manner as the nail covers the finger-end." But there can be no doubt that *ungulus* was merely the literal translation of the Greek ὀνύχιον, its derivation from *unguis* exactly corresponding to that of the original from ὄνυξ. The Greeks used the diminutive to express that the native substance had been modified to the purposes of art, according to a well-known rule of the language, of which examples are χρυσίον, ἀργύριον, gold, and silver, when coined. For the Tricoloured Agate being the usual signet-stone of the Greco-Italians and Sicilians, the Romans, when they began to obtain and to value the works of these engravers, at the same time adopted and Latinised the current designation of the material.

This explanation also throws light upon the apparently unintelligible definition of the Onyx extracted by Pliny from that very ancient mineralogist Sotacus, that " it exhibited the colours of the human nail, and also of the chry-

solite, the sard, and the jasper :" and again the statement
which follows, taken from Zenothemis, that the Indian
Onyx presented "many different colours; fiery, black, and
horny, surrounded by opaque white veins like an eye;" and,
further, that of Satyrus, that the same species was "flesh-
coloured, having one part of the chrysolite, another of the
amethyst" (xxxvii. 24). For these banded Agates, the
produce of India alone, cut across their layers, often present
the most beautiful and vivid colours, mimicking the Topaz,
the Garnet, and the Jacinth, in strongly-contrasted juxta-
position. Similarly the Sardonyx of the Greeks equally
differed in appearance, though not in species, from that of
the imperial Romans. Pliny quotes (xxxvii. 23) several
Greek authorities, to the effect that by the Indian Sardonyx
they understood a Sard with a white layer (candor in sarda),
and both colours transparent. The opaque kind, which in
Pliny's times he notes had engrossed the designation, they
termed the *Blind*, and apparently held it in no esteem.
Their Sardonyx was, in fact, the same evenly-stratified
union of Calcedony and Jasper as their ὀνύχιον; with the
distinction that its colours were white and red, not white
and black: for *red* was the proper and indispensable character
of the Sard. Pliny proceeds to give the reason for its universal use as a signet-stone amongst the Greeks upon its first
introduction from India: it was because the soft wax used
for sealing did not adhere to the engravings in this material, in which respect the Sardonyx excelled almost every
other gem. He uses the word "initio," *at first*, in evident
allusion to the total neglect of the gem in this capacity in
his own times. For amongst the purely Greek and Etruscan
works that Tricoloured Agate is frequent which presents its
bands in sard-red and white as well as in the dark-brown
and white of the original ὀνύχιον. This valuable property
of the gem, its non-adhesion to the wax, is a sufficient

reason why the Greeks should have adopted it for their signets, above all others, although its deeply-coloured and contrasted bands completely prevent the intaglio sunk in them from producing any effect to the eye. The signet, however, in those early ages was designed for use, not ornament.

No better definition of the Onyx can be given than in the words of Hill (' Translation of Theophrastus') : " The zones are laid with perfect regularity, and do not, in the judgment of the nicest distinguishers of the present times, exclude it from the Onyx class, of whatsoever colour they are except red ; in which case it takes the name of *Sardonyx*. The colour of the ground and the regularity of the zones are therefore the distinguishing characteristics of this stone; and in the last particularity it differs from the Agate, which often has the same colours, but placed in irregular clouds, veins, or spots."

Köhler, however, who has treated this question more fully than any other writer in his ' Untersuchung über den Sard, den Onyx, und den Sardonyx,' basing his explanation upon the numerous and conflicting definitions of the Onyx, extracted by Pliny from various Grecian authorities, has arrived at a conclusion differing very widely from Hill's : " The question, how is the Onyx to be distinguished from the Sardonyx, is now easily to be answered out of Pliny. As far as regards the substance and the colours, both are one and the same stone. It is called the *Onyx*, when the red, brown, or yellow ground is covered by white veins irregularly and capriciously disposed. If therefore these veins formed sometimes stripes, sometimes spots, sometimes eyes, then was the stone the Onyx. But if the various colours of the stone lay in regular strata one over the other, then it became the *Sardonyx*." That the concentric arrangement of the veins was the peculiar distinction of the Onyx, appears from Pliny's brief notice (xxxvii. 71) of its variety known as the Eye-onyx: "Triopthalmus cum Onyche nascitur, tres hominis oculos simul exprimens." The fact being that in Pliny's age the name no longer bore the same acceptation as when Theophrastus used it, but was certainly applied to the stone now called the Agate. This is the only way of understanding the passages already quoted from Sotacus and Zenothemis, as to its various colours surrounded by white veins like an eye, and these again sometimes traversed by other white veins running athwart them; and the remark of the former that " the true Onyx presents numerous and differently-coloured veins with milk-white zones, the transition of colours between them being perfectly indescribable, yet all blending into one harmonious whole, highly charming to the eye." It is singular, however, that Pliny should give no definition in his *own* terms, of what the Romans understood by the designation *Onyx*, but should have contented himself with citing these all but unintelligible descriptions of earlier and Grecian authors, being apparently himself in doubt as to the exact species of gem to which they referred.

Epiphanius, with his usual ignorance of his subject, confounds the marble with the gem Onyx: " Twelfth Stone (in the Rationale) the Onyx: this hath its colour exceedingly yellow (ξανθή). It is said that the wives of the kings (of Persia) and of the nobles take great delight in this stone, and have it made into drinking-cups for their own use. Besides this, there are other sorts, similarly called *Onychites*, resembling yellow bee's-wax. And some pretend that such are formed by the condensation of water dropping. These are called Onychites (nail-stones) in natural history, because the finger-nail in well-bred persons is made up of two colours, a marble-white coupled with the appearance of the blood underneath showing through the substance of the nail. Some, however, occasionally term marble Onychites, from the mode of testing it (by the finger-nail), or from the purity of its whiteness; but they are in error."

One passage, however, aided by tradition, will afford a clue for tracing the exact meaning of the name amongst the Romans, at least as far as regards the Arabian species. For this is described by Sotacus as differing from the others, in being black with zones of opaque white, whereas the Indian exhibited fiery spots encircled by transparent zones, either one or many around each, but differing from the same marks in the Indian Sardonyx "for in the former these spots are a shade (momentum), in the latter an actual circle." The only conclusion to be drawn from this observation is (with Köhler) that Sotacus called the irregularly stratified stone (our Agate) the Onyx, the regularly stratified the Sardonyx. Now the ring-stone most in favour with the Imperial Romans, next to the fiery Sard, if we may judge from the large number remaining, and the high character of the work they generally present, is a stone of two horizontal layers, the lowest of them black, sometimes opaque, but often tawny by transmitted light, covered by another extremely thin (Pliny's *momentum*) of milky-white, which, from the reflexion of the dark ground underneath, mostly shows like a turquois-blue. Another Nicolo exhibits the same colours more vividly in consequence of the total opacity of its substance. This arrangement of colours was procured by cutting out one of the "eyes" above mentioned, together with its encircling zone, and then carefully reducing it to the form best calculated to preserve the exact distribution of the two shades, by sloping off the sides but leaving the surface perfectly plane. The popular name for such a gem amongst the Romans was *Ægyptilla*, allusive to its origin, or rather the place of exportation:

"Nomen a loco: vulgus in nigra radice cærulea facie."
But the older definition, quoted from Iacchus, shows the
name originally designated the Sardonyx cut across the
layers, for he makes the stone distinguished by veins
(bands) of Sard and black passing across a transparent white
ground, " per album sardæ nigræque venis transeuntibus"
(xxxvii. 54). Köhler supposes, with good reason, that the
first-mentioned two-coloured gem is Pliny's Arabian Sardonyx, " which retained no trace of the Sard," for if the
third and uppermost layer of the Sardonyx be removed, the
remaining two will give us the very characters in question.
This is the Nicolo of the Italians, a corruption of *Onicolo*,
" a little Onyx," still called by the Germans *Onykel :* a traditionary name affording in itself a strong testimony that
this was anciently the Onyx *par éminence ;* and, as a diminutive, presenting a curious analogy with the original name
ὀνύχιον. A most preposterous derivation for the modern
title is indeed current, that the stone was so called after
a certain famous (and fabulous) artist Nicolò, who worked
in it in preference to any other; equally ridiculous with
that deduced by old Camillo from *Nicolaus*, signifying in Greek the " conqueror of nations," in virtue of
which designation the gem makes the wearer victorious in
battle!   A remark of the Pesarese Esculapius nevertheless
valuable as testifying to the existence of the name Nicolo
for this gem amongst the Italians of the 15th century.
This variety, or more correctly speaking, this method of
cutting the stone and giving it its characteristic appearance, was unknown in Grecian art, the intagli occurring in
it being invariably in the Roman manner, and in style all
posterior to the reign of Nero.   Infinitely the most splendid
specimen both for natural quality and for art that has
come under my notice was Hertz's, the *gem* of his cabinet,
an oval nearly two inches high, of the richest blue and

black, engraved in a very bold manner with Apollo resting
his lyre on a column, and standing before a tall, smoking,
tripod. It realised the high price of 90*l.* at the sale in
1859, and was bought by Mr. John Webb.

As an additional proof, derived from unbroken tradition
("traditione continua"), that the Nicolo was the Roman
Onyx, we have the curious fact recorded by De Boot
(writing on immemorial authority): "That sort of the
Onyx is the first in value and estimation which is of a
bluish tinge, having a black layer underneath; and is
especially sought after by the Jews, for they have a constant tradition that this bluish Onyx was one of the number
of the Twelve Stones, and therefore they reverence and
value it highly (ii. 244).

It has likewise the honour of being the first amongst
gems to be mentioned in the most ancient of all records
(Gen. ii. 12). Together with fine gold and the spice
Bdellium it was the production for which the land of Havilah was famed. The region intended by this name some
suppose to have been the Northern provinces of Arabia,
where Strabo places the *Chavilatæi.* Others translate *Bdellium* by "Pearl," in which case the passage contains the
three most noted exports of Arabia.

But to return to the definitions of the early Greek
travellers. Such a species of many-coloured quartz as
they describe, partly opaque, partly transparent, full of
lines, eyes, and spots, would now be called an Agate, rather
than an Onyx. Indeed Isidorus (Origg. xvi. 11), by a
singular anticipation of the modern nomenclature, describes
the *Achates* as "black, having in the middle white and
black circles joined together and veined." Such a description exactly applies to that finest specimen of antique work
in Agate anywhere preserved, the leopard's head of the
Townley Collection, admirably sculptured in full relief,

and as large as the head of an actual tom-cat: * the surface
marbled with shades of the purest black, white, and blue,
forming spots and meanders fully verifying Pliny's expression, " colore inennarrabili et in unum redeunte concentum
suavitate grata." But in fact the Agate and the Onyx are,
as already remarked, the same substance, only differing in
the arrangement of the layers, which in the Agate are
wavy and often concentric, whilst in the Onyx they are
placed parallel to each other.

Agathe-onyx is the name appropriately given by the
French to that variety in which the upper layer is opaque
and white, the lower transparent, and either colourless or a
pale yellow. This is the material most frequently employed for modern camei, being the production of Europe,
and often termed the German Onyx, whereas the ancients
preferred almost exclusively for that purpose the opaque
and rich-coloured strata of the Indian Sardonyx. Reversing the rule with modern camei, in the antique the relief is
usually rendered in pearly white upon a *black* ground.
There is also a substance closely resembling the Agathe-
onyx in appearance, though much softer when tested by the
file, and often to be met with employed in Italian camei
of the Renaissance, the *Albâtre-onyx* of the French. This is
nothing better than a variety of the Oriental Alabaster,
with the difference that the opaque and the transparent
layers, instead of intermingling in circles and curls, are
arranged in well-defined horizontal layers. The Italians
have also skilfully forged the antique Onyx by cutting out
a relief in the white lava from Monte Ipomeo, so much

---

* The Opal eyes, which impart such life and ferocity, were, says
Blumenbach, inserted by J. Pichler. The story current in the British
Museum is that this head capped Tippoo Sahib's tent-pole; a tradition
quite at variance with Blumenbach's notice, for J. Pichler died in 1791,
and Seringapatam was not taken before May 4, 1799.

used for Neapolitan ornaments, and fusing this down by
means of an enamel upon a ground of brown Agate; a
deception extremely difficult to detect. Brückmann mentions a singular variety of the Agathe-onyx in his own
cabinet; an antique cameo, where the upper layer was calcareous, effervescing with acid, the lower a pure transparent
quartz.

The most important works of the ancients remaining to
us in Onyx, are their vases of different forms, which have
ever been prized as the choicest ornament of imperial and
royal treasuries: enjoying, if possible, a higher estimation
in mediæval times than amongst their original possessors.
The earliest notice extant of Onyx vases occurs in Appian
(Bell. Mith. 115), where he enumerates amongst the treasures of Mithridates captured at Talaura, 2000 vessels of
Onyx (λίθου ὀνυχιτίδος), that is, of the *gem*, not marble, as the
feminine gender of the noun proves (indicating according
to the genius of the language, a *precious* stone), as well as
the notice that these vases were in gold mountings. This
seems to be the same collection with that which Posidonius
describes as " Onyx bowls, found in *nests*, or sets, up to the
capacity of two cotylæ (nearly one pint)." Their comparatively small dimensions also serve to prove that the
precious Onyx, equally with the Agate Murrhina, is here
intended, for in the *alabaster* so-called vases of the most
gigantic measurement were easily procurable, as the
splendid productions of the Volterra fabrique amply manifest in our own times. Epiphanius, as quoted above, speaks
of the Oriental princesses as delighting in drinking-vessels
cut out of the Onyx. The treasures of Mithridates came
to him in part from the inheritance of his ancestor Darius
Hystaspis, in part had been amassed by the Ptolemies, and
recently given up to the Pontic king by the Coans, in
whose safe keeping they had been deposited by Cleopatra

Selene. The greater porportion, however, was due to his own collecting, which he had prosecuted with unremitting zeal.

Veltheim ('Onyx-gebirge,' p. 75) correctly supposes that such vases were cut in nodules (Nieren), made up of concentric layers of Jasper and Calcedony, like the well-known Agate nodules. The ancient artist, adapting the form of his vase with admirable skill to the natural disposition of the strata in the stone under his hands, obtained a white coating enveloping the coloured body of the vase, through which he cut down to the dark field, thus producing designs in cameo. As works of art, these vases possess considerable merit; but infinitely more extraordinary are they, regarded as mineralogical specimens, as will appear upon the consideration of the dimensions of the principal examples preserved. Of these, the most famous is the " Cup of the Ptolemies," a *carchesium*, or two-handled vase, holding a sextarius (above a pint), being 4⅟ inches high, by 15⅟ in circumference, measured over the handles. It is covered with masks, vases set out upon a table under a vine from whose branches depend *oscilla*, and other Bacchic emblems, admirably executed in relief ; and hence its popular appellation, from the assumption based upon these embellishments, that it could have been executed for no other than Ptolemy XI., surnamed Dionysos. The style, however, is much more that of the times of Nero, a great dilettante, as Pliny records, in murrhine and crystal vases ; in fact, on one of his coins (in small brass), the reverse presents a table supporting similar articles, and altogether recalling the ornamentation of this carchesium. By a singular coincidence, we find here a Dionysiac utensil dedicated to the service of a St. Dionysius, and reverting to its original destination. For after its presentation in the ninth century, by Charles III. (the Simple) to the abbey of

St. Denys, it was thenceforward used, according to the tradition preserved by its historian, Dom Doublet, to hold the consecrated wine at the coronation of the queens of France. Its gold mounting bore a legend, added at the time of its dedication :

> " Hoc vas Christe tibi devota mente sacravit
> Tertius in Franco sublimis regmine Carlus."

The chalice was stolen in 1804 from the Musée, and its gold mounting enriched with gems melted down by the thieves: fortunately they were arrested in Holland, and the vase recovered undamaged ; and it has been elegantly remounted by Delafontaine. One fact shows the high value formerly set upon this relic: Henri II. pawned it to the Jews of Metz for a million of livres tournois (50,000*l*.), equivalent to five times that amount in modern currency.

Next in importance amongst these relics of imperial splendour comes the Farnese Tazza, a *patera*, or flat bowl, cut out of a single piece, eight inches wide by two deep, and carved with a subject in the Egyptian taste, allusive to the influence of the Nile. The land, personified by Isis reclining on her sphinx, holding aloft her wheat-sheaf, reposes under the protection of Father Nilus, and his two daughters of the Delta : the winds, authors of his rise, hover overhead. It is said to have been discovered in the *penetralia* of Hadrian's Mausoleum, when explored in the 16th century. It was purchased for 10,000 ducats, and has long adorned the Museum at Naples—a memento of the Farnese Princes of Parma, ancestors of the Neapolitan Bourbons.

Still more famous from its history is the Mantuan, or Brunswick Vase, an *alabastron*, or tall perfume jar, with narrow neck, five inches high by two in the greatest diameter. The relief upon it refers to the Eleusinian Mysteries, representing Ceres with Triptolemus in her

(G)   Q

dragon-borne car, three votaresses with their offerings, and the exergue filled up with masks and other Bacchic symbols. The style indicates, in the opinion of the most competent judges, the age of the Antonines, the middle of the second century. It has been tastefully fitted in the Cinque-cento mode, with a base and spout to convert it into a rose-water ewer, according to the usage of that period. It had long adorned the noble Gonzagu Gallery at Mantua, until, in the lamentable four days' sack of that city by the Imperialists in 1629, it fell into the hands of Fr. Albert, Duke of Saxony, one of the captains of the German robbers. The value formerly set upon it was 150,000 thalers, half which amount is said to have been frequently offered for it to its then possessor. The Duke's widow, Christina Margaretha of Mecklenhurgh, left it to her sister, Sophia of Brunswick, whence its present appellation. At Berlin is the " Beutisch Vase," the subject upon which Sillig interprets as commemorating the birth of Augustus; but Hirt, on better grounds, that of Commodus. Into the Russian Cabinet has found its way the very elegant one, formerly belonging to Caylus, and fully described by him (Rec. II. pl. 76). In shape it resembles the Portland, and is sculptured with two groups —Apollo seated, Cupid, and Psyche bound ; and Amorini chasing butterflies. Its height is three inches ; its material an Agate of three strata. The Vienna Cabinet possesses another alabastron, similar in shape to the Mantuan, covered with reliefs of masks and Bacchic ornaments ; but bearing for dedicatory inscription, a verse of Anacreon's, showing it to have been made for a present to a lady of pleasure :
" Ζήσαις ἐν ἀγαθοῖς, φίλη γὰρ εἰ ξένοις, ἔασον δέ με διψῶντα πιεῖν."

Other alabastra in Onyx, though unadorned with camei, were long treasured in the French cathedrals, and devoted to their original purpose of holding (the consecrated) oil.

A magnificent specimen of the stone, formerly belonging
to the cathedral of Sens, is now in the possession of Mr. J.
Webb. Amongst our Henry III.'s gatherings for his projected shrine of the Confessor, are put down " Phiola
oniclea et alia crystallina."

Such alabastra were occasionally diverted from their
proper purpose and used by the Romans for wine-flasks, in
order to impart their perfumed flavour to the liquor;
Martial notes (xiv. 110), " Ampulla Potoria "—

> " Hac licet in gemma quæ servat nomina Cosmi *
> Luxuriose bibas si foliata sitis."

The Onyx is enumerated amongst the articles of export
from India early in the first century by the author of the
' Periplus of the Red Sea.' Thus he mentions " the city
called Ozene (now Ougein), whence all things necessary to
the natives are brought down to Barygaza (Broach, Baroach,
Baroche, on the Gulf of Cambay), and also articles pertaining to the trade with us, namely, stones, the *Onyx* and
Murrhina, Indian cloths—fine and common quality, &c."
Again he has : " from the same places are exported nard,
zedoary, assafœtida, ivory, Onyx-stones, &c. To Barace
(Barcelore) are brought pepper, many fine stones; also
various and numerous kinds of lustrous gems, the diamond,
the hyacinthus, &c." Further on he specifies the exact
locality producing the Onyx :† " From Plythanæ (Pultanah)
Onyx-stones in abundance, from Tagaræ (Dowlatabad) much

---

* The most fashionable perfumer of the day.

† Colonel Stirling, the only modern I can discover who has seen
these stones in their native regions, speaks of the abundance of " the
Agates and Calcedonies of the Nerbudda, and last, but not least, the
great variety of the Onyx-stones of Malwa, above the Vindhyas. The
fields and eminences there are in many places spotted grey with these
last, from the size of a man's hand downwards. Many of them I have
had cut and polished, and they are the admiration of all ' admirers.' "

common cloths, &c., all which are brought down to Bary-gaza." Under the Ptolemies, the Indians themselves brought their own commodities to the city Eudæmon on the Red Sea, where they met the Egyptian traders as at a common entrepôt, like Alexandria in later times. But this place, notes our author, had been recently destroyed by Cæsar. Two centuries and a half later, we find from Vopiscus ('Aurelian,' 3), Firmus, a merchant of Alexandria, who usurped the people in Aurelian's reign, having ships of his own trading direct with India. Scythicus also, to whom Epiphanius ascribes the true authorship of the heresy of Manes, was an Arabian trading regularly between Alexandria and the Indian coast.

Having at length obtained an account of the present state of the Onyx-mines in the " land of Havilah," the Malwah district, its great interest both to antiquary and naturalist will more than justify its insertion here :—

" From Ruttunpoor we proceeded to the mines, distant one and a half *kos*, through a thick jungle. The whole road and all the *nalas* were strewed with Agates; but these are not of a good description, and are therefore not used. To the left of the road is a high hill, covered with jungle, and on the summit there is a *peer's* tomb. The only people residing there are a few *sidees* or negroes, who say that they were born and bred on the spot whither their fathers came from Broach. The mines that are now worked are situated on the sloping side of a small hill, covered with jungle, and extend upwards of four miles. At the time there were upwards of one thousand men at work, chiefly Coolies and Mussulmans. Each man collects a *maund* and a half of good stones daily. The shafts of the mines are about four feet in diameter, so that the miners in going up and down do not require the assistance of ropes, &c. They cut niches in the sides of the shafts for their toes to rest upon, and by pressing their backs firmly against the sides they in this manner ascend and descend without danger or difficulty. This I myself accomplished easily after going down two or three mines.

" The average depth of the shafts is thirty feet. I descended into one that was thirty-eight feet. The galleries run off in every direction, or wherever the miner's fancy leads him to dig. Their height is five feet and their width about four. The roof is arched, and the soil is a stiff clay in which the stones are embedded. The galleries seldom extend more than one hundred yards in length; but many of them join those of other mines. To each of the mines there are thirteen men attached, who work by turns; each man must send up so many basketfuls of earth and stones, when he is relieved. All the other people are employed in sorting and trying the stones as they are brought up. They seat themselves around the mouth of the shaft and try each stone, which is done by chipping off a piece with another stone. From the appearance of the fracture they are able to judge whether the stone is good or not; the finer and more compact the stone the better it will be when burned, and the blacker it appears at first the redder it will become after undergoing the same operation and when ready for being cut and polished.

" The stones are brought up by means of a rude roller or pulley supported by four pieces of wood let into the ground. This, with a small iron pick, not steeled in order that the miner may not be injured by the sparks, a few bamboo baskets, and one rope, compose all their implements. In one mine they had tapped a spring of fresh water at thirty feet, and had been obliged to abandon it until the hot season, when they stated they should be able to work it again.

" The strata through which the shafts are sunk appear

to be all nearly alike. The superficial bed consists of gravel, then red and yellow ochre, below which fuller's earth and ochre again; then a thin seam of rock containing a large proportion of iron, below which lies the clay in which the Carnelians are embedded.

" Each miner on his return to Ruttunpoor in the evening carries a basketful of good stones, when they are spread out on the ground and exposed to the sun. They are thus collected for a whole year, and turned every four or five days: the longer they are exposed to the sun the deeper or brighter the colour becomes when the stone is polished.

" In the month of May they undergo the process of burning, which is effected by placing the stones in black earthen *chatties* or pots. The pot is placed with the mouth downwards and a hole is broken in the bottom; a piece of broken pot is then placed over the hole, and the whole is then covered with sheeps'-dung, as no other material is said to answer for fuel in this operation. The pots are arranged in single rows, and the fires, which are always lighted at sunset, are allowed to burn till sunrise, when the pots are examined, and should there appear any white spots on the surface of the *chatty*, it is considered that the stones are not sufficiently burned, and they are allowed to remain for a short time longer. After this process the stones are all re-examined: those that have flaws, &c., are thrown aside, and those that are not sufficiently burned are laid by for next year's burning; the remainder are sold for exportation.

" The revenue derived from the mines is very insignificant at present; but with proper management I should say it might amount to something considerable, if we may judge from the custom derived by the Nawaub of Cambay, where almost all the stones are cut and polished and pay a heavy duty both on importation and exportation. Very few, if any, stones are now cut at Broach.

"It may be proper to add here a description of the process of cutting, polishing, &c., the stones, as carried on at Cambay in the present day.

"The most extensive consumption of Agates is required for the manufacture of beads, of every variety of size and colour, which are in demand for exportation to the African and Arabian coasts, as also for the islands of Zanzibar and Mozambique, where they are bartered for ivory, gold-dust, rhinoceros'-horns, &c. The stones are in the first instance broken into small pieces by means of hammers, and the beads are formed in the rough in the following manner :— An iron pin is driven into the ground with the sharp point upwards. The person operating places the part of the stone he wishes to break on this point, and with a small hammer he strikes the stone and continues the process until it has become partially round. This operation is performed chiefly by women, boys, and young girls. The work is then carried to another person, who proceeds with the operation. He has a large slab of hard sandstone before him, placed in a sloping or inclined position, and with a *clam* made of two pieces of wood, with a joint at one end, with two nails in the centre by which the stone is held ; he works the stone over the surface of the slab, constantly changing its position, so that in a very short time it becomes round. After this it goes to the polisher and then to the driller. The hole is drilled by means of diamond-dust and water. The drill is supported on a small frame and worked by a long bow backwards and forwards. Seals, &c., are all cut in this way, except those that require flat surfaces. This is effected by a *lap*, made of coarse lac and *koorun* (country emery), and formed into a thin wheel, two feet in diameter, supported on two pivots, and worked by a strap of leather pulled backwards and forwards. Bloodstones, Agates, &c., are formed into various articles by means of sheet-iron wheels of various diameters, worked as the others, and supplied with diamond-dust and water. The Carnelian beads are finished by putting a number into a bag in which they are shaken together."—(' A Visit, December, 1832, to the Carnelian Mines, situated in the Rajpeepta Hills, to the eastward of Broach. By Lieutenant G. Fulljames.')

Both the Onyx and the Sardonyx are now largely imitated by artificially preparing the German species, a stratified quartz, in quality little superior to common flint, by which treatment factitious stones are produced of much beauty, exhibiting well-defined and even layers of black and white, or red and white, equally clear and vivid.* This is effected by boiling the stones continuously, for the space of three weeks, in a solution of honey in water, constantly replenished as it evaporates, and afterwards steeping them in sulphuric acid to bring out the black and white; in nitric for the red and white colours.† The more porous strata absorb the honey first, and, secondly, the acid that carbonizes it—the more compact remaining unaffected by either. Such imitative gems are of very trifling value, being imported in vast quantities from Germany, where the secret was either discovered some years ago, or else, as others assert, introduced from Italy, where it had been practised from time immemorial. Pliny himself notes that *all gems* are brightened up by boiling in honey (mellis decoctu nitescunt), especially in the Corsican, noted for its acridity, although all other acids are detrimental to them (xxxvii. 74). There

* The true antique Sardonyx has a peculiarity that serves, in addition to the superior beauty of its colours, to distinguish it infallibly from these German counterfeits. Its black substratum when viewed by transmitted light, turns to a rich chocolate-red highly translucent, whereas the same part of the artificially stained recent species continues as black and opaque as before.

† Another sort display shades of a beautiful sapphirine produced by a similar application of the prussiate of potash.

is another very curious passage in Pliny, in the same
chapter, bearing directly upon the use of honey in the
treatment of gems :—" The Cochlides,* now a very common
gem, is rather an artificial than a natural production. It
is said to be found in Arabia in large masses ; and after a
boiling of seven days and nights continuously in honey,
in order to purge away the earthy dross, the residuum is
arranged, according to the fancy of the artist, in veins,
lines, and spots of various colours, in the manner most
likely to attract purchasers. Of such a size have these
gems been produced as to serve for frontlets for the horses
of Eastern kings, and for pendants to their trappings."
What variegated mineral substance is meant by this description it is impossible even to conjecture. It may after
all be merely a jeweller's story coined to account for the
whimsical conformation of the veins and shades in the
Agate ; or still more in the wonderfully beautiful streaky
pastes created by the Alexandrian glass-workers. The
name " Cochlides" indeed signifies " snail-shell stone ;"
and such is the literal meaning of the Italian *Lumachella*, a
dusky marble full of fragments of nautilus-shells, flame
coloured and glowing as if actually in a blaze. It is therefore possible that both terms, the ancient and the modern,
apply to the same substance, which from its beauty as well
as singularity must ever have been admired as soon as
known.

It is true that certain species of the antique gems, such
as the best Sards and frequent examples of the Onyx and
Sardonyx, are incomparably superior to anything of the
kind that is to be met with in nature at the present day ;
but it would be presumptuous to ascribe this excellence to
any artificial treatment of the subject-matter by the ancient

* This stone formed, says Trebellius Pollio, the clasp of Zenobia's
zone.

lapidary. Much more reason is there to suppose it the result of the incalculably better and more copious supply flowing in from a variety of sources now closed to us, such as Arabia, continually referred to by Pliny as producing the choicest gems, and the interior of Asia. Their Roman appellations inform us that this was the case with many of the ancient fine-coloured marbles, only known at present as existing in the remains of Cæsarian magnificence; such as the Rosso, the Nero, and the Giallo Antico, all from Numidia; the Paonazzo or Synnadic, from Phrygia; the Thebaic, from Egypt, &c. Numidia and Ethiopia, perpetually quoted by Pliny as exporting marbles and gems of all kinds, are now completely lost to the trade, as also are the equally productive mines anciently skirting the Red Sea. In fact, these were the regions supplying the Greek gem-dealers before India was opened out to them; as Theophrastus distinctly states (34): " First-class (precious) stones are rare, and come from only a few places; such as Carthage and the country around Massilia (for amber and coral, doubtless), and out of Egypt, from Syene near the city of Elephantina, and from the district called Psephò, . . . and the stones used for inlaying-work, out of Bactria."

To return to the artificial colouring of gems, it is ascertained by experiment that those belonging to any species of translucent quartz, which have been rendered white and opaque by the action of fire, will in a great degree recover their original colours upon immersion in hot water. This fact is strikingly exemplified in a tricoloured Agate belonging to myself, of which the two darker bands are completely calcined and chalk-like, yet these, after a short soaking, again become of two shades of brown, and as translucent as ever. Winckelmann had noticed the same transformation in a stone ('Pierres Gravées,' No. 1123), engraved with an Apollo, which from white and opaque became brown and transparent after having been worn a few hours upon the finger, in consequence of the absorption of the perspiration from the skin by its porous substance. Not being aware of the cause, he supposed the gem to be a Hydrophane or Oculus Mundi, a kind of Opal that goes through a similar change if saturated with water. Similarly another gem of mine, a fine Sard, engraved with a head of Sapor II., filled with white spots owing to a partial calcination, recovers its pristine pure and crimson transparency after a few days' wear on the finger. These experiments tend to verify Pliny's recipe for improving the lustre of the Carthaginian Carbunculus (above quoted), by a fortnight's maceration in vinegar.

Ben Mansur describes the Onyx (Dschesi) as divided into two species:—The *Bakrawi* of three layers: the first, opaque red; the next white; the third, transparent like crystal: and the *Habeschi*, also of three layers—two dark, and a white one between them (the actual ὀνύχιον of the Greeks). "Some are striped, others are not: in others the stripes are so interrupted that they form curious figures." * Interesting also is his notice of the localities yielding it :— " Although the Onyx is found in many places, yet those most valued are dug up on the confines of India and of China." From which it appears that, like the early Greeks, the Persian mineralogist makes no distinction between the Onyx and the Sardonyx.

In the same measure as the Alexandrian glass-workers counterfeited with astonishing success the natural clouds and polished surface of the Murrhina, so did they reproduce with equal skill, in their cheaper material, the precious Onyx vases and the cameo designs embellishing their surface. This was effected by fusing an opaque white

* On which account it takes with us the popular name of " Fortification-Agate."

coating over the coloured body of the vase ; and by cutting
this away where required, the design stood out in relief,
usually upon a dark-blue ground.   The process, as the
marks of the graver show, was precisely that employed in
cutting camei in hard stones.   Fragments of large vessels
thus adorned are frequently met with amongst ancient
ruins.   Of the perfect the Portland Vase is the most famous
specimen ; and next to this the Neapolitan, covered with
an elegant arabesque of interlaced vine-branches (VITRUM,
p. 332).   These ambitious attempts are aptly termed by
Martial, "audacis plebeia toreumata vitri,"—*toreuma* being
the technical name for ornamentation in relief, especially
when executed upon the surface of vases, whether in metal
or in stone.   Another epigram (xiv. 115) informs us that
the seat of this manufacture was Alexandria, and that the
risk of breakage during the operation was very great,
as might well be concluded from the fragility of the
material : —

> " Aspicis ingenium Nili quibus addere plura
>   Dum cupit, ah ! quoties perdidit auctor opus."

Yet the greatest artists of the day did not disdain to
exercise their taste upon so valueless a medium, for the
relievi on the vases above mentioned are far superior both
in design and execution to any decorating the celebrated
vases carved in the real Onyx.   This application of the
Glyptic art appears to be referable to Nero's times, as may
be deduced from Pliny's observation (xxxvi. 66), that a
peculiar mode of working glass (arte vitri) had then been
discovered which made two small cups of the kind called
pterotæ, or two-handled, sell for 6000 sesterces (60*l.*).   So
large a price could only have been commanded by their
artistic excellence, and not for any novelty in the colour or
body of the material; for it is mentioned in the same

way with the prices quoted repeatedly by the same author (xxx. 35) as paid for silver chasings by the old and eminent *cælatores.* Similarly Apuleius, in describing the banquet given by the wealthy Byrrhena, enumerates " ampli calices variæ quidem gratiæ, sed pretiositatis unius. Hic, vitrum fabre sigillatum, ibi crystallum impunctum, argentum alibi clarum atque aurum fulgurans, et succinum mire cavatum in capides ut bibas." Here it will be observed he makes the glass vessels to be adorned with figures in *relief,* the crystal with such in *intaglio*—a distinction confirmed by all the specimens extant in either kind.

De Boot states that the species composed of black and white layers was commonly forged in his times in order to be sold for a real cameo, in the following curious manner (ii. 94) :—" Take the little sea-shells used by the Italian women for paint (pearl-powder ?), grind them fine ; then steep the powder in lemon-juice, frequently passed through a filter, so that the juice be three or four fingers deep over it. Let the mixture stand thus, well covered over, for ten days in a warm place ; then pour off the juice ; wash the residuum in water ; grind it up in a porphyry mortar with white of egg well beaten up beforehand ; then cast the cement in moulds of the required design, made in wax. Next polish with great nicety the rough back of these casts, so that they may be applied skilfully and neatly upon a black ground, that the deception may not be readily detected. In rubbing down the shells, other colours can be added in fine powder, so that the Sardonyx may be imitated in this way as well as the other precious stones."

Georgius Agricola (who died in 1485) has a passage interesting as showing how rapidly the newly-revived art of gem-engraving had found its way into Germany :— " Nowadays, in the opaque white crust of the German Onyx (Agathe-onyx), engravers cut the coats-of-arms on which *our* nobles pride themselves; because this stone has veins more translucent than any other of the same kind, and moreover is harder than rock-crystal.   They also paint the *back* of these coats-of-arms with the proper *tinctures* required by the armorial bearings."

The Onyx, strange to say, considering its high repute in ancient times, bore a most unfavourable character in the Middle Ages, Marbodus asserting that its wearer was exposed to the assaults of demons and of ugly visions by night, besides being pestered with quarrels and lawsuits by day.   The sole remedy was to wear a Sard at the same time, which would completely neutralise the mischievous influence of the Onyx.

OPALUS: 'Οπάλλιον: *Opal.*

THE appropriate and forcible similes illustrating Pliny's poetical description of his Opalus sufficiently declare that he is speaking of the beautiful gem still bearing that name, and of no other. " Made up of the glories of the most precious gems, to describe it is a matter of inexpressible difficulty : there is in it the gentler fire of the Ruby, there is the brilliant purple of the Amethyst, there is the sea-green of the Emerald, all shining together in an incredible union. Some aim at rivalling in lustre the brightest azure (Armenium) of the painter's palette, others the flame of burning sulphur, or of a fire quickened by oil."

This charming iridescence is not however inherent in the stone, but produced by the reflexion and refraction of the light imprisoned within certain openings in the interior of its mass, which are not fissures, but arranged in regular directions. Newton ascertained by experiment that inclosed plates, or *laminæ*, of air produced according to their thickness the several prismatic colours after certain definite proportions; and from this rule he derives the cause of this beautiful property of the Opal. It is composed of pure Silica and water, the latter imbibed from the atmosphere. This structure renders it particularly fragile and tender; sometimes even flying into pieces on a sudden exposure to either extreme of cold or of heat. It is subject also to the loss of its splendour, when the laminated openings on which

the play of colours depends become choked up with dust
and grease through wear; although it is said this property
may be revived by the hazardous operation of roasting the
stone by a gradual heat. The Hungarian Opals exhibiting
a uniform milkiness of surface, more or less iridescent,
have from their greater density the advantage of resisting
the effects of wear longer than any other sort; hence their
superior value. But infinitely greater is the beauty of the
Mexican when recent, presenting an unmixed globule of
green fire like the glowworm's lamp, or a ball of phos-
phorus moistened with oil. Nevertheless of so porous a
nature is this kind, that it becomes colourless if wetted;
and, however carefully preserved, changes to an opaque
brown after a brief existence as a jewel, and consequently
it has no value in the gem-market. Barbot warns us
against the dealer's artifice of macerating inferior stones in
oil (to be detected by the tongue) which forces out their
colours in a transitory perfection.

Pliny justly observes (xxxvii 21), "the Opal differs at
once very much and very little from the Beryl," referring,
it would appear, to the glassy texture of its substance,
common to both, yet with the grand distinction of its
iridescence. It came next to the Emerald in value: and
naturally so, being almost the only one of the really precious stones of which the full perfection could be developed,
by the simple process of polishing known to the Romans.
For this stone is invariably cut *en cabochon ;* and owing to
its extreme softness, the polishing is a matter of time and
care alone. This is the sole instance where Pliny has
quoted the selling price of a precious stone : the famous
Opal of Nonius, then still in existence, as large as a hazel-
nut (the greatest magnitude known up to that time to be
attained in the class), valued at 20,000*l.* of our money
(vicies H. S.). Nonius was proscribed by M. Antony

purely for the sake of this gem ; but made his escape, carrying off the ring along with him, the sole relic of his fortune : putting up with exile rather than make his peace with the Triumvir by the surrender of the coveted treasure. The finest Opal of modern times was the Empress Josephine's, entitled the " Burning of Troy," from the innumerable red flames blazing upon its surface, the reverse being perfectly opaque. The present owner of this unique gem is unknown. The Turks of our day esteem the Opal as highly as the Diamond, and readily give 1000*l.* for a perfect one of the size above specified by Pliny.

" India," continues the naturalist, " is the sole mother of the Opal." This region (or perhaps Arabia) continued the only source of the supply of the best sort down to a comparatively recent period. When De Boot wrote there was but one mine known in Hungary, and that nearly filled up with rubbish ; and which besides, he remarks, had hardly ever furnished anything but the third and fourth rate Opals, the first even then coming exclusively from India.* Some mineralogists doubt the fact that any region of the East Indies ever produced the true Opal, merely because no such gem is now brought from thence ; but the same argument applies here as in the case of the true Emerald, not at this time found in that country, formerly the principal source of the stone. The Hungarian mines of Czernovitza (where this gem occurs embedded in porphyry), a region inaccessible to the Romans of Pliny's age, now supply the finest Opals, and of a size far exceeding that of Nonius. A gigantic specimen is exhibited in the Imperial Cabinet at Vienna, and for which,

---

* The defect of the Hungarian was the yellow tinge pervading the body, which entirely dulled the precious iridescence of the species. The great majority too were no more than Senei-Opal (the Girasol).

(G)   R

according to report, the incredible sum of 50,000*l.* has been offered and refused.*

" Some," says Pliny, " have given to this stone, on account of its pre-eminent loveliness, the name of the *Pœderos,* or Cupid." Others made the Pæderos a distinct species, called *Sangenon* by the Indians; but produced in many and different localities, Egypt, Arabia, most abundantly in Pontus, in Galatia, Thasos, and Cyprus. Of this the best sort presented somewhat of the beauty of the Opal, though seldom free from flaws (scaber); but its colour was made up exclusively of purple and sky-blue (aëre), the emerald-green was entirely wanting. In this species, " to be overcharged with a wine-colour was preferable to the being too pale and watery "—terms, these, which show that by *Pœderos* was understood the bright Indian Amethyst, in parts almost colourless, in parts clouded with the richest purple, and often exhibiting a slight iridescence in the flaws to which its body is too subject. It was a stone so admired by the Romans, that, employing it always cut in a bossy cabochon, they placed but the most minute intaglio upon its centre, to detract as little as possible from the native beauty of the material. In fact, Pliny (40), after stating that the perfection of the Amethyst consisted in a rose-colour, as it were borrowed from the Ruby, shining mildly amidst the purple, and most striking when the gem is held up against the light (*in suspectu*), a property very observable in the lighter Almandines, adds, " that such gems some prefer to designate as the Pædcros,

---

* Caire was convinced of the existence of the *Indian* Opal, having met with some indubitably Indian cut; and which when tested on the wheel proved considerably harder than the Hungarian sort. The Turks, however, are fully persuaded that this beautiful gem is due to no earthly mine, but falls from heaven in the lightning, thus holding it for the veritable Ceraunia.

or the Anteros, others as the gem of Venus." Again (46),
he describes the Pæderos as " the chief amongst the *colourless* stones (candidarum dux), though it was a question
under what *colour* it ought to be classed : the name having
been so much bandied about amongst the beauties of other
species, so that the mere distinction of loveliness had
become of itself a name." Meaning that this name of
Cupid had been indiscriminately applied to the most
elegant specimens of stones of many different sorts, provided they possessed the qualification of *rosiness*, being in
fact a mere epithet of beauty, not of species. " Nevertheless, the actual kind (or the Opal specially so designated)
comes up to our expectations of what is due to such a name.
For in a transparent Crystal are united a sky-blue, turning
to a peculiar green (viridis suo modo aër) a gleam of purple
(our crimson), and also a certain golden, vinous hue ;
always the last in sight, and always crowned by the purple.
The gem appears steeped in these colours singly, and yet
in all at once : no other is more limpid or agreeable to the
eye. The best kind is found in India, where it is called
Sangenon; the next in Egypt, termed there Tenites; the
third is the Arabian." A passage this, leading to the
same conclusion as the one previously quoted, *viz.*, that the
Pæderos properly signified the Opal, though applied by
some lapidaries to the Amethyst and pinkish Almandine.

The defects of his true Opal are thus enumerated : its
colour passed off into that of the flower called *heliotropium,*
or into that of crystal or of hail ; sometimes intermixed
with *salt* (white granulations), or flaws, or little points that
struck the eye. The semi-Opal is merely a pure calcedony
of a pale, milky blue, which in certain lights appears to
contain a slight fiery radiance. From its *deficiency* in the
proper beauties of the Opal the Germans styled it the

Waise, that is, the Orphan. But with the Italians the same stone went by the name of *Girasole* in virtue of its radiance when turned towards the solar beams.

No other gem was so successfully imitated by a glass paste (similitudine indiscreta): the only test for detecting such being to hold them in the sun, when, if the false gems were poised upon the finger and thumb against his rays, a single colour showed through the mass unchanged, and was spent within itself. In the real Opal, on the contrary, the colour perpetually changed, shooting forth now beams of one, now of another tinge, and diffusing their lustre over the fingers supporting it. Nothing shows the wonderful skill of the ancient glass-workers more convincingly than this testimony of Pliny's to their success in counterfeiting the Opal, a gem of which it has hitherto baffled the skill of the Parisian paste fabricant, to produce an imitation at all capable of deceiving the connoisseur.*

From its then enormous value, as well as on account of its fragile nature, the Opal must have been rarely submitted to the skill of the Roman engraver; for the earlier Greeks were totally unacquainted with the gem. Hence Professor Urlichs justly pronounces unique the Opal of the (former) Praun Collection, engraved with the head of Sol between those of Jupiter and Luna. The somewhat debased style shows it to be a work of the Lower Empire, when the all-prevailing faith in astrology had sufficient

---

* It is, however, possible that Pliny here alludes to the Girasol Calcedony, which greatly resembles a bad Hungarian Opal, exhibiting a flamy reflexion, but unmixed with green, and which the Romans doubtless considered as an Opal. This kind is indeed exactly and easily counterfeited in glass. Of something of the kind (or else of *schmelze*) were the glass opalescent cups (allassontes) sent by Hadrian to Servianus as specimens of the Alexandrian manufacture.

influence to induce the possessor of so precious a medium to convert it into an amulet, of virtue doubtless commensurate with its price. Another magnificent Opal, though corroded by time, is set in a cabalistic ring of the 13th century, now in the Braybrooke Collection.

A stone apparently so instinct with life, naturally bore an abundant crop of superstitions to the vivid imaginations of the multitude in the Dark Ages. Some slight traces of these are already apparent in Orpheus (279), who declares that the Opal, "displaying the complexion of a lovely boy" (alluding to its epithet Pæderos), gives delight to the Immortals, and is also a protection to the sight. Marbodus, converting the word into *Ophthalmius*, improves largely upon the simple dictum of his predecessor, adding that it also confers the gift of invisibility upon the wearer, so that the thief protected by it could carry off his plunder in open day.* Still lower down in the Middle Ages the Opal was believed to possess united the special virtue of every gem with whose distinctive colour it was emblazoned. Even De Boot evidently believes "that the wearing of it preserves the sight and the brightness of the eyes." But the climax is attained in the extravagant laudations delivered by Petrus Arlensis, a visionary of Henri IV.'s reign. "The various colours in the Opal tend greatly to

---

* *Opalus* was supposed to be only another form of *Ophthalmius*, 'Eye-stone," whence sprung these notions of its virtue. This derivation gave birth to the old spelling "Ophal:" take, for example, in the list of Queen Elizabeth's jewels (Harl. MSS.), "A Flower of gold garnished with sparkes of Diamonds, Rubies, and *Ophals*, with an Agate of her Majestie's viznomy; with a pearle pendant; and devices painted upon it. Given by eight Masks in the Christmas week anno regni 24ᵐᵒ." Some conjecture the Opal to be the gem called by the rabbins the "Calf's-eye," which, however, was much more probably a three-coloured Onyx, like that known to the Greeks as the Lycophthalmus.

the delectation of the sight; nay, more, they have the very greatest efficacy in cheering the heart and inward parts, and specially rejoice the beholder's eyes. One in particular came into my hands, in which such beauty, loveliness, and grace shone forth, that it could truly boast that it drew all gems to itself; while it surprised, astonished, and held captive, without escape or intermission, the hearts of all that beheld it. It was of the size of a hazel-nut, and grasped in the claws of a golden eagle, wrought with wonderful skill. It had such vivid and various colours, that all the beauty of the heavens might be viewed within it. Grace went out from it; majesty shot forth from its almost divine splendour. It sent forth such bright and piercing beams, that it struck terror into all beholders. In a word, it conferred upon the wearer all the qualities granted by nature to itself; for by an invisible dart it penetrated the soul and dazzled the eyes of all who saw it, appalling all hearts however bold and courageous. In a word, it filled with trembling the bodies of the bystanders, and forced them by a predestinated impulse to love, honour, and worship it. I have seen, I have felt it, I call God to witness! Of a truth, such a stone is to be valued at an inestimable amount." By a strange freak of superstition the Opal has lost this, its ancient glory. Barbot says, " Certain groundless stories, founded, doubtless, upon the legend of Robert the Devil, have in our day discouraged the use of this gem as an ornament. People accuse it (and this in the nineteenth century!) of bringing ill-luck upon the wearer. It were useless to point out the absurdity of this supposed malignant influence, which is manifested, as they say, by the fading of its colours : a change really due to the atmospheric and accidental causes already pointed out."

The mode of cutting this very fragile stone is first upon a leaden plate coated with *adoucis* (emery-powder that has been already used upon other gems, and so *deprived of its asperity*); next on a wooden wheel with fine pumice-powder; finally on a third covered with felt moistened. The last polish is given with a cloth and tripoli.

The Musée de Minéralogie, Paris, possesses a very fine Opal carved into a bust of Louis XIII., when a child—a piece of unreasoning extravagance, where work and material mutually destroy each other's beauty and value : the iridescence entirely neutralising all effect of light and shade in the relief.

OSTRACIAS, or OSTRACITIS : *Crude Vitriol?*
Pyrites?

THUS briefly noticed by Pliny (65) :—" Like a potsherd
(testacea), of a harder quality than the Ceramitis, resembling the Agate, except that the Agate becomes smooth
and fatty by polishing. The harder kind possesses such
force that gems may be engraved with its fragments. The
Ostracitis derives its name from its similarity to a shell
(or oyster)." Two distinct substances are here described
as the same, the confusion arising from the circumstance
that ὄστρακον means both a potsherd and an oyster-shell.
All to be learnt is that the *Ostracias* was some mineral
capable of being used in gem-engraving, resembling burnt
clay in colour, rough, and incapable of polish ; the *Ostracitis*
merely a fossil shell.* What the Ostracias really was may,
however, be *conjectured* from the other notices of it. " The
Cadmitis is the same as what is called Ostracitis, except
that the last is sometimes surrounded by azure bubbles
(bullæ)." The *Cadmea*, possibly so named after the mythic
Phœnician father of mining in Greece, was either copper
or tin ore, being "ipse lapis ex quo fit æs" (xxxiv. 22).
But it also signified certain chemical products of its fusion
adhering to the sides of the furnace, and therefore of a
vitriolic and arsenical nature, much used in the ancient

---

* Pliny seems to have included these minerals in his list of *Gems*,
merely in virtue of their employment in engraving.

pharmacopœia as caustics. Of these the most ponderous, the " Placitis," was divided into the " Onychitis," " externally azure, but spotted like the Onyx inside;" and the " Ostracias," "entirely black," and the most powerful of all. It may therefore have been an agent in gem-engraving, like vitriol at present.

Cadmia (or Cadmea) is supposed to mean also Calaminestone, from which one may suspect that the Ostracias of the Greek gem-engravers was the Marcasite, or Iron Pyrites, which, fractured, much resembles Calamine-stone; when roasted, turns to a rusty red (testacea), and which has been used from time immemorial in Persia for polishing the harder gems. This last is indeed the strongest of my reasons for their identification. Ben Mansur states, " the Laal (Spinel) takes a polish with difficulty, and for a long time they were unable to polish it, until at last they brought it about by means of the Gold Marcasite called *Ebrendsche.*" De Laet, in 1647, states that the Spinel and Balais could only be polished by means of the Pyrites. Modern lapidaries use vitriol for the same purpose, the principle of both processes being identical, vitriol being merely the extract of the Iron Pyrites.

It is indubitable that the Greeks derived all their processes of the Glyptic art from Persia, whence by the way of the Ionian cities the use of intagli had been introduced amongst them; and whence, from Sardis in particular, most of the gems they possessed were exported.

But the use of Pyrites amongst the Romans in this art is not a mere matter of conjecture. Heraclius, in his all but unintelligible doggrel, ' De Artibus Romanorum,' written in the 7th century, actually speaks of engraving upon *glass* with "the hard stone known by the name of Pyrites."

Nevertheless my conjecture that actual copper-slag was

meant by *Ostracias* is favoured by a passage in the same hard-to-be-understood author, where he describes the polishing of Rock-Crystal by means of "fornacis fragmine, micas," ground to powder, and spread upon a leaden plate,—

> " Hæc etenim plumbum conjunctio reddit acutum,
> Et suum rursus habent *lateris* fragmenta vigorem."

Yet it seems hardly possible that any merely vitrified matter should be sufficiently hard to be thus employed as a substitute for emery, even though " fornacis fragmen " means the Ostracias of the copper-furnace.

Again, if we suppose the name Ostracitis to be derived from the resemblance to a shell, it is a singular coincidence that fossil shells, especially echini and small ammonites, are very plentful, composed entirely of Iron Pyrites, forming beautiful golden objects which must have often excited the admiration of the ancient miners.[*]

From its use to the lapidary the Marcasite was an important article with the Persians. Ben Mansur devotes a separate chapter to it, dividing the kinds, according to their colour, into gold, silver, iron, and copper Marcasites. Marcasites, when facetted, have a true adamantine lustre (though perfectly opaque), which they preserve without tarnishing. In the last century they were much used in jewelry as a cheap substitute for Diamonds. Pyrites, from their supposed fiery nature (they were the equivalent for flint in the tinder-box of the Greeks and Italians, and in the early wheel-lock guns of the latter) were esteemed powerful agents in the reducing of tumours; being roasted, ground, and made into a plaster. Dioscorides and Galen give recipes for their employment.

---

[*] Which Pliny actually records of the latter fossil. "Ammon's Horn, reckoned among the most holy gems of Ethiopia, is of a *golden colour*, and presents the image of a ram's horn : it is pretended that it procures prophetic dreams " (60).

OVUM ANGUINUM: 'Οφίτης: *Adder-bead:*
Serpentine-marble.

" MOREOVER, there is a kind of egg in mighty reputation in
the Gallic provinces, of which Greek writers have no mention. Innumerable snakes, twining together in summer,
make a ball, by a skilful combination, out of the froth from
their jaws and the slime of their bodies. It is called the
Ovum Anguinum. The Druids assert that it is tossed on
high by the serpents' breath, and must be caught in a cloak
before it touch the ground: the robber makes his escape
on horseback, the snakes following him till they are
stopped by the intervention of a running water. The test
of its reality is, that it should float against the stream, even
though set in gold; and, so ingenious are magicians in
disguising their impositions, they declare it must (to have
any virtue) be captured at a particular age of the moon—
as though it were within the will of man that such an
operation should coincide with any determined time. I
have myself certainly seen the egg, which is of the size
of a small, round apple, covered with a cartilaginous crust,
with many excrescences, like the suckers on the arms of
the cuttle-fish. It was being worn at the moment as the
badge of a Druid. It is marvellously extolled for its effect
in giving success in battle and in petitions to princes: a
proof of the falsity of which is the fact, that to my knowledge a Roman knight from Vocontii was put to death by.

the Emperor Claudius for no other cause than the carrying of one in his bosom during a trial. This embracing together of the serpents, and their thus productive union, seems to be the cause why foreign nations have made the Caduceus entwined with snakes, one of the emblems of peace; for it is not the custom to represent the snakes on the Caduceus as having crests" (*i. e.* not as the serpents called by the Romans "dracones," the sacred reptiles above all others, the Egyptian Agathodæmons).

These details show how much in the wrong antiquaries are in giving the name of Ovum Anguinum, or Druid's Bead, to the large spheres of variegated glass (often exhibiting the most vivid colours, arranged in elegant wavy patterns equal to any traditionally reproduced by the Murano factories) that formed the central ornament of Roman-British or Saxon necklaces. But the object seen by Pliny was evidently some natural production, and his description applies better to a large fossil echinus than to anything else. Perhaps some lingering traces of the ancient fable may be the cause why a species of the recent shell yet bears the popular name of the Mermaid's Egg. This explanation too is corroborated by Mediæval tradition; for De Boot actually figures two fossil echini as specimens of what in his day was accounted the true Ovum Anguinum, and prized as an antidote against poison; but he sagaciously adds that they do not agree with Pliny's description. The longitudinal septa of the shell were construed into as many lizards attached to its surface, and are so depicted in his woodcut.

Singularly enough he notices that the travelling quacks were in the habit of selling to the simple Bohemians, as the veritable Serpent's Egg, and generated according to all the circumstances of the old Druidical legend, a thing hence called *Duchanek,* or the Breath-stone (lapis Spiritalis).

This was a thick glass ring, bluish green in colour, having but a small opening. On the exterior circumference were set equidistant projections in blue and white, like eyes. De Boot warns his readers against being taken in by these mysterious-looking objects, stating that in Belgium in his boyhood (about 1555) they were commonly used by women as an ornamental counterpoise to the end of the spindle ("spindle-whorls"), for which purpose they were manufactured there. These identical articles occasionally are turned up in England,* and are still received without question as indubitable Druid's Beads.

In the Middle Ages this wonderful production passed for no more than an antidote against poison, pestilence, bad air, &c.; but we see from Pliny's notice that anciently the virtues attributed to it were of a much higher and supernatural order—not merely medicinal. It was for the attempt to pervert justice by magical practice that Claudius, " the wisest fool, or the foolishest wise man " (as was said of our James I.), amongst the emperors, put to death the Gallic gentleman who had come armed with one into Court, his cause apparently being tried before the Emperor himself: in which lay the peculiar heinousness of the offence.

The extraordinary appearance of these fossils, often filled up with gold-like pyrites, and inclosed in the heart of large masses of chalk, or lying deep in the solid clay (*gault*), necessarily rendered them objects of wonder, and perhaps of religious awe, to their ignorant finders, in the days when geology was an unknown science. Even common pebbles, of a more than usually perfect spherical form, seem to have been regarded by the Celts as holy things. It is said that some have been preserved on the chapel-altars (like

---

* One was shown me found several years back at Llanturnam, Monmouthshire, and highly valued by the owner as a true Druidical relic.

the *Salagrama* of the Hindoos) in remote Irish districts from time immemorial, and the water in which they are steeped considered a specific for diseases in cattle. The head of a certain Highland clan also inherits, by descent from the remotest ages, a white sphere possessing similar virtues, in all probability transmitted to him from the Druidical period through an endless line of Gaelic ancestry.

This famous Gallic talisman bears a singular analogy, both in its name and properties, to the Ophites or " Serpent-stone " of the Asiatic Greeks ; for Pliny's *Ophites* (xxxvi. 11) has no connexion with our subject, being only the green serpentine-marble used in his time in decorative architecture. But Orpheus (355) styles it (the gift of Apollo to Helenus) " the vocal stone, the truth-telling Sideritis, which some mortals prefer to call the Ophites, in which dwells a soul (ἔμψυχος), round, rough, black, hard. All over its circumference run sinews like unto wrinkles, deeply scored upon its surface." After keeping a fast, and perfect chastity for thrice seven days, the seer bathed the stone in running water, clothed it in soft raiment like an infant, and setting it up in an illuminated shrine, inspired its substance with life by means of certain mighty spells chanted over it. Then taking it in his arms he fondled it like as a mother does her babe, whilst he questioned the spirit within, and received true responses to all his demands. By this means did he reveal to the Atridæ how his native city Troy was capable of being captured. But Orpheus warns whoever prepares to consult the oracle, to arm himself with courage, lest, when he first hears the stone, grow ing instinct with life, utter a cry like that of a new-born babe desiring the breast, his hands unnerved by terror let the ball drop upon the ground, and thus excite the heavy wrath of heaven. But if he had courage enough to consult the oracle, it would return to every question the most veridical answers. At the end, the consulting party is ordered to bathe the stone a second time, and holding it to his eyes watch it closely, when he will perceive the spirit departing out of it in a wondrous manner.

This last character suggests the idea of the Hydrophane, a brown opaque kind of Opal, which, after immersion in water, becomes transparent and lustrous, returning to its primitive dull opacity as the liquid evaporates from its pores. The gem on this account was greatly admired in former times, and was honoured with the title of *Oculus Mundi*.

Another virtue of the Ophites alleged by Orpheus (413) is that the bearer may walk securely among serpents, though they should meet him in swarms. If even he tread upon them with bare feet they never think of using their teeth, but either hasten away terror-struck, or else, if rushing to the attack, stop midway in their career, and, raising their necks on high, fawn upon him with their tongues like so many dogs. Guarded by this amulet, the hunter Euphorbus slept in the forests of Ida unhurt, close by the most savage dragons. The powder of it also was a sure antidote for the serpent's bite. Besides this it was a cure for blindness, deafness, barrenness, and impotence. The *Salagrama*, or sacred stone of Vishnu, so much used by the Brahmins in all propitiatory rites, especially at the moment of death, is described by Sonnerat as a kind of *Ammonite*, very *heavy*, oval, or round, *black:* the actual wrinkled, ponderous, round, black Orites of Orpheus. Sometimes its colour is violet, which, representing a vindictive avatar of the god, is looked upon with horror. It resembles touchstone, and is concave on the reverse, with spiral lines terminating towards the centre. It is hollow, with only a single small aperture. The possessor of one keeps it wrapped up in a linen cloth, often perfumes and bathes it, and the water thus used is highly valued for its sin-repelling properties. Colebrook again describes the stone as round and black, perforated with spiral holes as if by worms, done by Vishnu's own finger, the forms and number of which typify the god under different characters. The substance is silicious. They are found in the Gandaci river in Nepal. The agreement in colour, weight, and spiral arrangement of the "deep-furrowed lines," proves the Orites and this stone to have been identical. Observe, it is given to Helenus by *Apollo*, who answers to the Hindoo Crishna, the most splendid of Vishnu's avatars. The Magi may easily have got the stone from the Brahmins along with so many other particulars of their religion.

The *Goa-stone* was in the 16th and 17th centuries as much in repute as the Bezoar, and for its similar virtues. It got the name from being brought by the Portuguese from their East Indian colony. It is of the shape and size of a duck's egg, has a greyish metallic lustre, and, though hard, is friable. The mode of employing it was to take a minute dose of the powder scraped from it in one's drink every morning, when wonderful results were promised in the preservation of the health. The substance is evidently a metallic compound, and somewhat resembles calomel in the mass. So precious was it esteemed that the great usually carried it about with them inclosed in a case of gold filigree.

The Memphitic kind of the *marble* Ophites, which approached in fineness of quality to a precious stone (gemmantis naturæ), was employed by surgeons as an anæsthetic agent. Powdered and mixed with vinegar, the paste was smeared over the parts that had to be cauterised or amputated. "By this application the body loses all pain, becoming as it were benumbed" (Plin. xxxvi. 11). An attempt has lately been made to revive the use of local anæsthetics in modern surgery.

Pliny accurately distinguishes his Ophites marble from the Lacedemonian (Verde Antique), " of a very precious kind, and the most gay of all (hilarius)." A similar marble, named after Augustus and Tiberius, had been discovered in Egypt during their respective reigns. This serpentine Marble is a compound of Magnesia and Silica in nearly əqual proportions, owing its colours to a small admixture əf Iron oxide, and of Manganese, and Chrome. When recent it is easily cut with the steel tool, but attains considerable hardness after exposure to the air, as is demonstrated by the excellent preservation of the plaques so plentifully introduced into the external inlaid decorative work of the Tuscan Gothic buildings. The Verde Antique has angular patches of pure white marble dispersed throughout an unmixed green mass, much resembling Plasma in its colour and slight translucency. The Ophites (Green Serpentine), on the contrary, is dull, opaque, and merely speckled with white, like the skin of the asp that gives the name. Pliny makes two kinds: the pale and softer, the dark of superior hardness. He notes that it did not afford columns, except of very small dimensions. In fact existing remains show that the Romans used it chiefly in veneers for coating the walls of rooms, or for inlaying in their "pavimenta sectilia." The ancients evidently knew nothing of the quarries at Prato near Florence, which now furnish Green Serpentine, " Verde di Prato," in inexhaustible quantities, and employed by the Tuscan marble-workers for vases, inkstands, and similar artistic manufactures in great variety.

The Red Serpentine of late years extensively quarried about the Lizard Point, Cornwall, and much used for the purpose of internal decorations, and for vases, is far superior to the Green in beauty. Its ground is red elegantly mottled with black, and it takes a fine and durable polish. Although necessarily, from the fact to be stated presently, known to the ancients, it was not reckoned a variety of the Ophites, but, as it would seem, classed amongst the Agates. Its sanguine hue would have caused it to fall under the denomination of *the Hæmachates.* This with the Green and the Black can justly claim the precedence in antiquity over all the other materials of the Glyptic art. They were almost exclusively employed for the cylinder signets belonging to the first period of Assyrian art, that is prior to the reign of Sargon (Shalmanesar), and the designs cut upon them are identical in character with the earliest bas-reliefs discovered at Nineveh. The Black is of a particularly fine and uniform texture, so that it commonly passes, in these remains, for actual black Jasper.

PANTARBES.

PLINY has no mention of this marvellous gem, but later
writers have more than made up for the omission. Apollonius Tyaneus, according to Philostratus, saw it during
his visit to India, and brought back the following account
(iii. 46). It was as big as the thumb-nail, of a fiery colour
and luminous by night. It was generated at four cubits'
depth underground; but so forcible was the exhalation
from it as to cause the superincumbent earth to crack and
thus betray its existence. It eluded the touch of the
vulgar, and could only be drawn forth by means of certain
rites and charms known to the Brahmins alone. But its
peculiar virtue was that of attracting all other precious
stones; for if any number of the latter were dropped into
a river or even into the sea, and this gem were let down
attached to a string, they all clustered about it like a
swarm of bees around their queen, no matter how far
dispersed at first, and were drawn up adhering to it.

Such a gem naturally was invested with the most
wonderful virtues of another kind. By means of such an
amulet Chariclea escapes unharmed from the pyre to which
she had been condemned by the jealous Arsace (Æthiop.
viii. 11); inasmuch as she had secreted about her the
espousal-ring of King Hydaspes, "which was set with the
stone called Pantarbes, engraved with certain sacred letters,
embodying, as it has proved, some divine charms, by means
of which a virtue is imparted to the gem antagonistic to

s 2

Fire; and thus giving to the wearer immunity from hurt in the very midst of the flames."

The fable recorded by Philostratus bears a strong resemblance to the mode of discovering the Topazius mentioned by Diodorus Siculus; and possibly refers to the same stone. The Indian name, modified by the Greek into Pantarbes (as formed of πᾶν and ταρβεῖν, in the sense of All-feared), seems to have suggested the notion of the deference paid to it by all other gems to which it stood in the relation of a queen-bee to her swarm.

It would seem that the Ruby is the stone intended by all these legends savouring so strongly of a Hindoo origin. The high estimation in which it was held, its fiery colour, and believed luminousness in the dark, are all points supporting this explanation. This also supplies the reason why this particular gem should have been selected for an amulet against the element of Fire; for the distinctive epithet of the true Carbunculus was *Acaustus*, "the incombustible:" and the substitution of an active for a passive virtue was facile with the Oriental worshippers of gems. Hafiz sings how that Jamshid's Ruby Cup blushed rosy like the dawn through the porphyry walls within which it was immured.

PORPHYRITES: *Porphyry.*

AN igneous rock of dark-crimson ground thickly dotted with small crystals of felspar. Though there are other colours of precisely the same stone, as far as its chemical composition goes, yet the ancient Porphyrites, "purple stone," designated exclusively the first-named kind. It took also the epithet Leptopsephos from its speckled texture. Egypt alone produced it in Pliny's age, and in masses of sufficient dimensions for the largest works. But under the Lower Empire the Romans obtained an inexhaustible supply of the finest quality from nearer home; Valery observed on the coast of Sardinia vast quarries of Porphyry with shafts of columns lying about, merely roughed out as they were left by the workmen. The earliest works in it seen at Rome were statues of Claudius brought from Egypt by his deputy Vitrasius Pollio, a novelty, which Pliny remarks, was not received with approbation, at least no one had up to that time followed Pollio's example.

Under the Lower Empire, however, it was largely employed in the most sumptuous edifices then erected, in the form of columns, of *labra* for the baths, and of sarcophagi. Some of these columns, in a single piece 42 ft. long, are still to be seen fulfilling their original destination (having mostly been destroyed for the sake of the material),* in

* Which was sawn up into thin plaques in Gothic times for making the "Opus Alexandrinum," the then much admired pavement for churches: in imitation of the ancient " sectile."

the portico to Constantine's Baptistery at Rome, and in the mosque of Sta. Sophia. It was a fasionable material for the lower parts of the later imperial busts, having the head alone in white marble or bronze, its colour aptly reproducing the sovereign purple (a dark crimson dye). Those ages have left to us certain works in Porphyry, the execution of which remains a mystery; perhaps the most wonderful things the Roman sculptor ever produced, considering the elaborateness and high relief of the decorations, and the invincibility of the material employed. The magnitude of the original blocks is likewise a subject for admiration. Of these monuments the principal are the Sarcophagus of the Empress Helena, in a single block, 19½ palms high by 12 long (13 × 8 ft.) of the finest texture and deepest colour. It is adorned on the two fronts with a military procession, the "Triumph of Constantine," comprising many figures on horse and foot. Busts of Helena and Constantine stand out in the highest relief, as medallions above. Placed at first in the Mausoleum of Helena (now the Torre Pignattara), it was removed thence to the Lateran by Anastatius IV., in 1154, to serve for his own tomb; and was set close to the Porta Santa until 1600, when it fell to pieces on being removed in the course of the alterations of that Basilica. The fragments remained in the cloister until restored and placed in the Vatican Gallery by Pius VII. The restoration and repolishing of this monument occupied many hands continuously during seven years.

The second, the Sarcophagus of Constantia, daughter of Constantine, is much less elaborate in its decorations, besides being of very inferior workmanship to the preceding. Cut like that out of a single block, 11 palms long by 8 high (7½ × 5½ ft.), the two fronts present groups, prettily designed, of Cupids engaged in the vintage : below

are peacocks and rams. On the ends stand boys holding
up bunches of grapes in allusion to Christ's words, "I am
the vine, ye are the branches." On account of these
sculptures it was known during the Middle Ages as "La
Tomba di Bacco," and remained uninjured on its original
site within the sepulchral chapel of Sta. Costanza, until
transferred by Pius VII. to the Vatican. A copy in
plaster occupies its ancient niche in the Mausoleum. And
to close the list, with the most important examples furnished
by modern times of the application of this costly material
to the honours of the sepulchre, the tomb wherein rests
the Duke of Wellington beneath the cupola of St. Paul's is
hewn out of one block of green Porphyry. The stone had
been prepared by nature for the purpose, at the beginning
of Time, in the shape of a boulder upon a Cornish moor.
Though perfectly devoid of ornament, the labour of two
entire years was required to reduce it to its pattern, *in situ.*
The Sarcophagus of Napoleon I., deposited in the sepulchral
chapel under the dome of the Invalides, is carved with a
grandiose simplicity of design from a single block of a red
stone, not a true Porphyry but a Quartz-gritstone, of even
harder texture. The rock was brought from the quarry of
Schorkisena, near St. Petersburgh, and cost, by the time it
reached its destination, no less than the equivalent to
5500*l.* The lid, also formed of one piece, is said to weigh
32 tons (?). Within lies the imperial corpse occupying
its original tin coffin, which is again enclosed in one of
mahogany, two of hard lead, and lastly in the sumptuous
case of ebony made for the occasion of the funereal procession on December 15th, 1840.

Small pieces of Porphyry, selected for their peculiarly
bright colour, were occasionally engraved upon by the
later Romans, but merely as talismanic intagli, and similar
devices making no pretensions to art. The ingenuity of

Renaissance gem-engravers, however, has availed itself of this substance: there is in the Galleria at Florence an excellent head of Leo X., an intaglio on a large circular seal, set in iron after the manner of a coin-die, and designed to impress the leaden bulls of that pontiff.

The art by which the Romans worked* these vast masses with such evident facility is entirely lost. At present, when made use of in architectural decoration, the only method of subduing the stone is to steep it for many weeks in urine (as a Roman architect who had had recent occasion to employ the stone informed me), and even then it speedily turns the edge of the best steel tools. It has therefore been conjectured that the old sculptors worked both this and Basalt by means of emery-powder and chisels of soft metal. It is known that the early Egyptians executed their stupendous works in hard stones with bronze chisels, some of which have been found in the quarries.

The Grand-Duke Cosimo I. was mightily vain of a discovery made by himself of how to temper steel tools by some chemical means so as to cut Porphyry (Galluzzi, iii. 127), and several statues executed by his order still adorn the entrance to the Boboli Gardens; but their execution is rough and clumsy, not approaching in the least to the beautiful finish of the ancient works in Porphyry.

---

* From its extensive dedication to the splendour of the old capital the Byzantines only knew Porphyry by the name of the ' Roman stone." Codinus quotes a letter of a noble lady, Marcia, accompanying the gift to Justinian for the fabric of Sta. Sophia, of certain columns so designated, stolen from Aurelian's Temple of the Sun.

PRASIUS: Πράσιος: *Plasma.*

THE ancient Prasius, so called from the resemblance of its colour to the peculiar green of the leek (πράσιον) is now confounded in the numerous class of green stones styled indiscriminately Plasma by the antiquary. This word, formerly written Prasma, whence the French name, Prisme d'Émeraude (the origin of the common, but deceptive designation *Root of Emerald*) is merely the Italian corruption of Prasina, according to their common vulgarism of changing R into L and vice versâ. Under the Lower Empire, when the slang of the circus had become the current language, Prasinus, the colour of the Green faction, came to denote that particular shade to the exclusion of the more general word Viridis. Hence we find the panegyrist (in Symmachus) using the word Prasini instead of Smaragdi; and Epiphanius ('XII. Stones of the Breastplate,' iii.) remarking, by a slight blunder, that the Smaragdus is called also the Prasius.

The Plasma, considered mineralogically, is merely translucent Calcedony coloured green by some metallic oxide, of copper when dark, of nickel when of a brighter shade. It is in fact a nearly transparent green Jasper, and though frequently approximating in colour to the finest Emerald, it is seldom quite pure, but marked with black spots, or yellow patches, betraying its proper species. In fact, some specimens are internally diversified with ramifications of a dark, opaque, green, so as to deserve the name of green-

Moss-Agates. There is, however, a pale-green variety, of which antique ring-stones are occasionally found, made perfectly free from all such blemishes and remarkably transparent. Such are specimens of the true Prasius of the ancients, their tint exhibiting an exact resemblance to the colour of a leek. These approximate closely to the true Emerald in their oily, soft tint, though they have not its peculiar lustre; and such were without doubt the Smaragdi, which the *Greeks* speak of as being engraved for signets. Pliny indicates the species exactly, by noticing that one kind was "horrent with drops of blood," our Heliotrope; and the third, marked by three opaque white lines. This last is the same stone as that described by him under "Jaspis" as universally worn for an amulet in the East; resembling the Emerald and surrounded by a white line passing through its middle, and therefore called the Grammatias; as that with more than· one line was termed Polygrammos. I have met with a Gnostic amulet on a stone exactly answering to this description of the Grammatias; and, conversely, by its means have identified the true Prase beyond all possibility of doubt.

But amongst the numerous varieties of the Plasma some are to be occasionally seen which distinguish themselves from the rest by the purity of their substance, and peculiar shade of light green slightly tinged with yellow. They strongly resemble the Prehnite (a combination of silica, alumina, lime), a stone as hard as the Garnet, and found in considerable botryoidal masses in the Tyrol as well as at the Cape of Good Hope. As we know the Romans obtained Rock-crystal from the Alps, there is no reason why they should not have discovered and made use of the Prehnite also: its colour and hardness would have highly recommended it to a people such lovers of green-coloured gems.

The commonness of the stone when Pliny wrote, is •
clearly shown by his expression "vilioris est turbæ
Prasius," "the Prase belongs to the vulgar herd." It was
extensively used for intagli by the Romans of the Lower
Empire, though but little at an earlier date, to judge from
the circumstance that although engravings in it are more
abundant than in any other material except the Sard and
Carnelian, yet any of really fine Roman work are excessively rare; and of the Greek period, hardly to be met
with. Indeed, the sole intaglio in Plasma possessing great
artistic merit, that has ever come under my notice, is a
bust of Severus in a large gem of remarkable beauty for
its purity and tint, lately acquired by the British Museum.
The subjects also engraved upon it are usually of the class
most in fashion in the times of the Decline, such as Eagles,
Victories, Venus and the Graces. It is strange that it
should not have been more employed by the artists of a
better period, both on account of its agreeable colour, and
its resemblance to Calcedony in the facility of working.
One would argue from this that the material was of late
importation into the Roman world, and, to support this
belief, another *good* work known to me in it represents a
girl, with her name added ΘΕѠΦΙΛΛ (perhaps the poetess
lauded by Martial), wearing her hair in the odd fashion
that came in under Domitian. As for camei in this stone,
though abundant enough, yet they seem with few exceptions to belong to the times of the Renaissance, or later.

The native country of the fine, antique Prase is now
unknown* (though an inferior sort is obtained in the
Faroe Isles), but lumps of it unworked are sometimes

---

* There is good reason for believing it India : for Dionysius Periegetes describes the natives of Ariania as collecting the "grass-green
Jaspis along with the Diamond, the Beryl, and other stones of price
from amongst the gravel of their torrents.

picked up amidst the débris of Roman edifices. Several of the green stones distinguished by Pliny with the names of Prasioides, Tanos, Molochites, and perhaps the Callaina, are, to all appearance, now included under the indiscriminate appellation of Plasma. Certain it is, that the great variety of the tints and textures of the stones so designated at present would have induced the ancients, whose scientific nomenclature was entirely based upon external peculiarities, to class them under different names.

Prismatical beads of Plasma, as well as of Garnet, are often to be discovered in the earth about Rome. They all range to nearly the same size, so that antiquaries have little difficulty in forming an even row out of many distinct purchases. An additional fact this tending to prove in its way, that our Plasma is but another name for the *Jaspis* of the ancients: for Naumachius names necklaces of the latter as equally coveted by women with those of the Hyacinthus (Sapphire).

Orpheus (749) sings: "By this means (by eating of the serpent-sacrifice) was I taught thy powerful virtue as a remedy against the sable asp, thou life-saving, divine stone, that bearest the name, and the colour, of the green leek." Here we find the root of the manifold virtues as an antidote ascribed to the Green Jasper by mediæval sages (JASPIS).

SANDASTER: *Aventurine* (?), *Matrix of Opal* (?).

A GEM so called from the locality in India that produced it; sometimes also known as the Garamantica, because coming from the interior of Africa. It was likewise found in Southern Arabia. It is described by Pliny as related to the Anthracitis (a substance like fossil charcoal, and probably Jet); but its value consisted in the circumstance that a fire shone forth inclosed within it, with golden drops glittering like stars, and these last always confined within the substance of the stone, not showing upon its exterior. (xxxvii. 28.)

Upon these data De Laet, and most mineralogists after him, pronounce the Sandaster to have been our Aventurine (then newly brought from India), a reddish-brown translucent quartz, filled with innumerable particles of gold-like mica. It takes its present name from the fact that the Venetian imitation of it, so often seen in Italian jewelry, as a ground for mosaics, and in earrings, brooches, &c., was found out *per aventura*, "by good luck;" from the accidental admixture of brass filings with melted glass.

But the innumerable specks of gold, the chief beauty of the Aventurine, prove of themselves that this was not the *Sandaster*: because the latter was held a sacred gem, as bearing an affinity to the heavenly bodies, on account of the stars within it being arranged as the constellation the Hyades stand in heaven, and in the same number—that is, only *five*. Besides, the golden specks in the Aventurine

would not have been described by Pliny as *stars*, implying
a certain definite form, but as *gold-dust*, the term he uses
for the similar appearance in the Lapis-lazuli.

An ingenious derivation for "Aventurine" has been
offered in the "Apanturin," *panther-stone*, a name given to
a species of Jasper in the Targum of Jonathan-Ben-Uzziel,
composed in the sixth century. But the "Pantheros"
is actually described by Marbodus, after some ancient
authority now lost, as coloured with black and white, red,
green, rose and purple; a character, totally unfitting the
aventurine quartz, but exactly applying to the Brocatella
Agate. It cannot be by an accidental coincidence that
Orpheus had before given these very same colours, in
spots, to the kind of Agate, the most esteemed for its protective virtues, which he calls the "Leontoseres," highly
extolled by the demi-gods of old. Now the specific virtue
assigned by Marbodus to his Pantheros is, that whosoever
looks upon it in the morning should be protected thereby
from all malice of his enemies throughout that day. In
fact, its influence was that of the mediæval St. Christopher :—

> " Christophori faciem die quocunque tueris
> Illo nempe die mala morte non morieris."

The brilliancy of the Indian Sandaster was so great as
to injure the sight, if viewed too long. Another character
quoted from Ismenias, that this gem could not be polished
in consequence of its tenderness, and therefore brought a
high price in its native condition, does not by any means
suit the Aventurine, a hard quartz-gem, and requiring a
polish to bring out its beauty. All the writers, consulted by the Roman naturalist, were agreed that the
more numerous the stars the gem contained, the greater
became its value.

Nothing now coming from India, or employed in jewelry, corresponds to this obscure account (Pliny himself had evidently never seen the stone) so well as the Matrix of Opal, a compact Serpentine of a dark brown colour, and filled with minute Opals of extraordinary lustre, mimicking indeed, to a fanciful eye, the stars that gem the ebon brow of night. The exceptional tenderness and fragility of the imprisoned Opals seems unmistakably referred to by the remark of Ismenias above cited. This beautiful substance cut into thin slabs, their value proportionate to the number of Opals they contained, was formerly much in fashion for the tops of snuff-boxes, and similar trinkets. The matrix itself being a decomposed serpentine is extremely fragile, and therefore is cut into such plaques with much difficulty; another point of resemblance to the above-given character of the Sandaster. Its natural colour is light yellow, but, before use, it is soaked in oil and baked, which besides rendering its texture more compact, turns it into a rich dark-brown, marvellously enhancing the lustre of the little Opal-stars twinkling over its surface.

The name *Sandaster* sounds like a Persian epithet, and appears to be corrupted from *Saan-Aster*, the "Royal Fire." Pliny notices that it was used by the Chaldeans in their religious offices, in virtue of the stars it contained. Another stone was often confounded with it from the similarity of its name merely, viz., the *Sandaresus*, or *Sandaresion*, also an Indian export. But this was of an apple or olive green (some species of the Jaspis therefore), and of no value.

The Lumachella marble has in one respect a strong claim to represent the Sandaster: its ground being dusky, and so far bearing out the comparison of the ancient mineral to the Anthracitis; but the fires blazing so wonderfully within its darkness assume the curved forms of the broken

shells to which their origin is due, and could not possibly
have represented *stars* to the imaginative Chaldeans, the
givers of its name.

But as India is designated the "sole mother" of the
precious Opal, the Hindoo lapidaries must have been
acquainted with the matrix also of the gem, and from its
enormous value have been inspired with the notion of
thus utilising its minutest particles under the high-sounding
title of " Sandaster."

SAPPHIRUS: Σάπφειρος: *Lapis-lazuli.*

THAT the Sapphirus of the ancients was our Lapis-lazuli, is evident from the remark of Theophrastus (23). " Of stones, there are also others out of which they engrave signets, some for the sake of their beauty alone, such as the Sard, the Jasper, and the Sapphirus: this last is, as it were, *spotted with gold-dust.*" The same appears from Pliny's more detailed description of it (xxxvii. 39), that it came from Media (whence the entire supply of Lapis-lazuli is brought to the present day, if we extend the name to Persia and Bokhara;)* that it was opaque and sprinkled with specks of gold ; of two sorts, some of a purple-tint (cum purpura), the best, and found in Media ; the others, of a dark blue colour (cyanei), were accounted the males of the species. Epiphanius is very instructive upon this head. " The Fifth Stone : the stone Sapphirus, purple in colour, like that of a dark-blue beetle. Of this there are many kinds : for there is the Royal spotted with gold, yet this is not so much esteemed as the sort altogether blue (unmixed). And this is reported to be found in India and Ethiopia, wherefore they pretend that the sacred place of Bacchus amongst the Indians has a flight of 365 steps made out of the Sapphirus :† though most people think

* Although plentifully found in China, and of late years in California, the produce of both these countries is too full of pyrites and white veins to be available for the Glyptic art, and is only good for calcination.

† Dionysius Periegetes describes the sandy deserts around Mount

(G) T

this story incredible. This gem is admirable and very
beautiful, and agreeable to the eye, on which account they
set it in bracelets and necklaces, especially for the wear
of princes. It is likewise medicinal, for being powdered
it heals the sores following pustules and boils if smeared
over them, being applied mixed with milk to the ulcera-
tions. It is written also in the Law that the vision seen
by Moses in the Mount, and the Law given unto him, were
made out of the stone Sapphirus" (*i. e.* were inscribed on
Tables of Lapis-lazuli).

Before the true precious stones were introduced from
India, the Lapis-lazuli held the highest place in the estimation of the primitive nations of Asia and Greece (as we
see from the terms in which Theophrastus describes it);
in fact, it was almost the only gem known to them having
beauty of colour to recommend it.

This again was the only stone of any intrinsic value
known to the Egyptians under the Pharaohs; hence it
abundantly occurs in their jewelry that has come down to
us, worked up into signet-tablets, pendants, and charms. But
this Egyptian sort appears to be of a very inferior vein, dull
and pale in colour. Still scarcer was the material amongst
the contemporary Assyrians, to judge from the rarity of
cylinders in Lapis-lazuli, though a few such do exist;
whence it may be concluded that the Egyptians had worked
mines of it in their own territories (the Ethiopian above
quoted) from the earliest period of their kingdom. This
stone had the popular name of *Syrtites*, says Isidorus,
because it was picked up in abundance upon the Syrtes or
quicksands of Cyrene on the African coast. There may be
some foundation for this story, for even now rolled masses

Parapamisus, the territory of the Ariani, as abounding with quarries
of the Sapphirus and *Coral.* This last is a manifest absurdity unless
he possibly means the Corallis, or the true antique Red Jasper.

of Lapis-lazuli together with Amber are washed up by heavy gales upon the coast of Capri, which lies opposite though remote.

Although Pliny says it was in his times considered unfit for engraving upon in consequence of its substance being full of crystalline points (the spots of pyrites that appear like gold), yet we have works in it of every period of antiquity. True it is that for such (when in a good style) have been selected portions of the pure unmixed blue, mentioned by Epiphanius as the sort most admired. These, too, have in many cases preserved their colour and original polish to an astonishing degree, proving the vastly superior hardness of this species over that known to the Egyptians. Greek intagli in this stone are extremely uncommon, yet a large scarabeoid recently brought from Athens (Rhodes), is engraved with a kneeling Venus robing herself, in the purest style of the age of Phidias. The Praun Cabinet boasted a noble head of some youthful king of Macedon, on a large circular disk of the deepest violet, the reverse also engraved with full-length figures of Apollo and Venus: a work pronounced by Steinbüchel contemporary with Alexander. The Blacas has Perseus, last of the line, wearing the winged helm of his mythic namesake. Both intagli and camei of the Roman times are frequent in this material, in spite of Pliny's assertion; though such are rarely works of great merit, yet fairly executed Roman relievi and intagli in Lapis-lazuli are often 'o be seen. Of the former the Medusa's Head formerly belonging to the Cav. Azara is always cited as the most remarkable for beauty and for magnitude. But of all antique works in Lapis-lazuli the most remarkable known to me is an owl about eight inches high cut out of the solid block with admirable vigour and truth to nature. Thus precious in its material there can be little doubt it originally accompanied a statuette in gold of the Athenian goddess, which graced some imperial *lararium,* perhaps Domitian's own. On the reverse of his fine gold medallion (Paris) his patron goddess appears with just such an owl as this at her feet. This unique piece of sculpture afterwards passed into the Cabinet of L. Fould. The Marlborough Cabinet possesses a few; more remarkable, however, for the extraordinary beauty and depth of colour of the material, than for the engravings (later Roman) upon them. As for camei of imperial date the finest known to me are a Messalina in front face (Marlborough), a Faustina in extremely high relief, and a Crispina in profile (both Praun). The Byzantines used it largely for the same purpose: a magnificent example in the Trésor de S. Denys, the donation as it would seem of the Emperor Heraclius, bore on one side the bust of the Madonna, on the other, of the Saviour. The Persians under the Sassanian dynasty often employed it for bearing regal portraits, witness the grand Varahran in the British Museum.

With the Italians of the Cinque-cento it was an especial favourite, particularly for vases (of which the Louvre Museum displays a matchless example,* the *nacelle* (boat) of extraordinary magnitude, valued in the inventory at 200,000 francs), and for miniature busts and small relievi.

A serious defect of this substance is that by exposure to heat and moisture it loses its beautiful azure, and assumes sometimes a black, sometimes a chalky appearance, yet the best quality used by the Romans has often retained both colour and polish unimpaired in wonderful perfection.

The Egyptians covered much of their small works in terra-cotta, idols, symbolical figures, rings, &c., with a

---

* This elegant appropriation of the stone to the purposes of art has lately been revived by the Parisian lapidaries with great success. The vases of M. Rudolphi vie in novelty and grace of form with those of the Renaissance.

coarse thick blue enamel, intended to give them the appearance of being carved out of this valuable mineral (CYANUS, p. 125). But the Romans, who carried the manufacture of pastes to such perfection as regards colour and hardness, were eminently successful in their production of an artificial Lapis-lazuli, hardly to be distinguished when polished from the genuine stone. The most wonderful example of the art ever produced is the Townley *Bonus Eventus* (British Museum), where the youthful deity is represented at three-quarter length and in half relief, in the most finished style of Roman art, upon a slab of this composition about eight inches square. The relief has been carefully worked over after leaving the matrix in the same manner as a cameo in hard stone; and nothing can be more perfect than the imitation of the azure and texture of the actual Lapis-lazuli. The figure is evidently the portrait of some youthful Cæsar, represented according to custom under the character of this deity of promise: the features, harsh and surly in their expression, though somewhat softened down by the artist, are indubitably those of the young Caracalla, thus inappropriately personified.

SARDIUS: Σάρδιον: Σάρδιος: *Oriental Carnelian.*

THE Sard, according to Pliny, is so called after the place Sardis, where it was first discovered. This etymology, however, is merely based upon the similarity of the words, and is of no more worth than the ludicrous derivation given by Epiphanius, from its resemblance in colour to a pickled sardine—an idea though that very naturally suggested itself to an icthyophagous Byzantine saint. But Σάρδιον, the earliest Greek form of the word, cannot be derived grammatically from Σάρδεις; its root is undoubtedly the Persian *Sered,* "yellowish red," very slightly altered. The name came with the thing from Persia; the Babylonian mine produced the sort first known and most esteemed; hence the stone is termed by Epiphanius the Babylonian Sard.

Of the modern name, Carnelian, the derivations are numerous, the usual one being assigned from its colour of raw flesh, *carneus.* Again, it is spelt Cornelian, as if from Corneolus, and equivalent to the German *Hornstein,* which last signifies the European sort. Lessing, with some plausibility, supposes this form taken from the French *Cornaline,* to describe its similarity in colour to the cornel cherry.*

The common Carnelian, a semi-transparent quartz co-

* The word *Carnelius* dates from the Decline, for we find Marbodus using it (xxii.) to denote a different and inferior stone to the Sardius. As he alludes to its colour of raw flesh, he evidently held to the first derivation of the name.

loured red or yellow by oxide of iron, and its superior
Oriental variety the Sard, hold the first place in the list of
substances employed by the ancient engravers, presenting
us, alone, with as many intagli cut upon them as all the
other species of gems collectively.

The Carnelian is found abundantly in many parts of
Europe, wherever the shingle on the coast is composed of
flint-pebbles, or in the beds of mountain-torrents of similar
formation, and scattered together with Agates over the
Egyptian Desert. It is of the same nature as the latter
stone, only differing in the arrangement of its colours, and
seems to be what Pliny distinguishes from the rest of the ·
species by the name of Sard-achates, just as his Leuc-achates
is the Calcedony, or White Carnelian.

In this dull red, earthy, and softer species are the most
ancient intagli usually cut, the Egyptian and Etruscan
scarabei, and the greater part of the other ring-stones
engraved in Etruria. The beds of the Tuscan rivers furnished a plentiful supply of this material; even at the
present day the shingle of the brook Mugnone,* near
Florence, yields Carnelians in great abundance. But the
beautiful transparent species, the true Sard, came from
India alone. Already (B.C. 400) Ctesias, in his ' Indica,'
mentions the " great mountains out of which are dug the
Sardò, the Onyx, and other gems," lying fifteen days'
journey from the sandy desert (between Cutch and Moultan);
and again, the " mount Sardò, and the mountains where the
gem Sardò is dug " (Ind. § 5). And Plato, with a traditional reminiscence of India in his mind (Phædo, p. 110),
describes the " True World " (Paradise) as a region where

---

* Whither the simpleton Calandrino, according to Boccaccio's tale
(viii. 3) goes with his tormentors Bruno and Buffalmacco in quest of the
gem Heliotrope, that was to give him the power of becoming invisible
at will.

all the rocks are that substance of which the gems so coveted by the Greeks—Sards, Jaspers, and Emeralds—were but fragments that had escaped the universal ruin of all things here below. But when the trade with the East was opened fully out by the conquests of Alexander, and the establishment of a powerful Greek kingdom in the North of India, the Sard came into general use. " No other stone," observes Pliny (xxxvii. 31), " was so great a favourite with the Greeks as this: at least the plays of Menander and of Philemon revel in allusions to it." On this stone nearly all the performances of the most celebrated antique artists are to be found, for as a general rule fine work was never thrown away upon an inferior or too obdurate material; and there was good cause for this preference, such are its toughness, facility in working, beauty of colour, and the high polish of which it is susceptible; which last, Pliny remarks, it retains longer than any other gem. The truth of this assertion has been confirmed by the eighteen centuries that have elapsed since he wrote, for antique Sards are found always retaining their original polish, unless where very roughly used; whilst harder gems—Garnets, Jacinths, and Nicoli— have their surfaces greatly scratched and roughened by wear. So true is this, that the existence of a perfect polish upon any one of the latter class affords in itself a tolerably sure proof that the engraving is either modern or has been retouched in modern times.

The gradations of colour exhibited by the antique Sard are almost innumerable. The bright cherry deepens into the fiery red of the Carbuncle, and thence into a semi-opaque black, only red when viewed by transmitted light. The bright pale yellow increases in intensity to the richest orange, and thence to a reddish-brown scarcely to be distinguished from the Jacinth. This again becomes overcharged with black till it darkens to the deepest coffee-colour, and complete opacity. In a rare variety a slight admixture of green produces an olive tint, but yet easily distinguishable from the Plasma. In all these, again, the variations of the intermediate shades are infinite. Of such varieties each obtained a distinctive name in the copious terminology of the ancients, and if the yellow, green, or brown prevailed, was ranged under the Jaspis. It is ignorance of the true meaning of the word that produces the modern epithets of " yellow " and " brown" *Sard.*

When Pliny wrote, the bright-red variety was the most esteemed; the honey-coloured (*i. e.* the yellow and semi-opaque) were of less value; but those of the colour of a burnt brick (testaceæ), or red and opaque (the common Carnelian), were utterly rejected. The bright-red are certainly very fine in tint; they are extremely transparent, and often come near to the Carbuncle in colour and in lustre, but may always be recognised by the tinge of yellow invariably mingled with their crimson. Those in which this colour deepens into the dark tint of the Morella cherry were considered the males of the species; for the Romans, following the Greek mineralogists, divided gems into males and females, according to the depth or lightness of their shades.* In this bright-red sort are the finest Roman intagli for the most part found. The light-yellow sort (Pliny's *Sphragides*, a class of the *Jaspis*), resembling amber, was much employed at an earlier period. On *this* most frequently occur the finest works of the Greek artists, more especially those stiffly-drawn but highly-finished figures of the most minute execution, surrounded with borders, which were formerly termed Etruscan, but now are with more reason ascribed to the Archaic Greek school. Some good Roman works occur in this variety, but they are few in number and of an early date, their scarcity confirming

* Theophrastus (30).

Pliny's statement as to the disrepute into which the yellow-coloured had fallen.

Bright-red, indeed, was the primary distinction of the Greek and Roman Sard, and hence the origin of Epiphanius' ridiculous etymology; and thus Marbodus characterizes it "Sardius est puniceus, cujus color sanguineus." But *now*, by a singular perversion of meaning, the French restrict the name *Sardoine* to the brownish yellow (*fauve*) calling the red alone *Cornaline*. The former word being naturally enough translated, by English and German writers, as *Sardonyx*, has given rise to much confusion in quoting from Catalogues drawn up in French. Similarly, the unaccountable practice of the English lapidaries of the last century, in calling *Beryls* the yellow shades of Sard, has proved to the French and German archæologists (like Clarac and Dr. Brunn) a fruitful source of error in their (copied) descriptions of engraved gems. This *Sardoine* runs to a very much larger size than the true *Sard*. Amongst the French Regalia one is catalogued as of 7⅜ inches in diameter, engraved with the Head of Medusa, and valued at 480*l*. The Marlborough "Phryne" (or rather a Venus Victrix) to be reckoned amongst the finest intagli of the Roman period is also a rich dark *Sardoine*, 2¼ inches high.

The supply of Sards from Babylon had failed before Pliny's age. They were said to have been found in the vicinity of that city, enclosed, like a heart, within other stones on their being broken open, and therefore termed "lapicidinæ." This means, speaking popularly, that the Sard nodules formed the "plums" in a pudding-stone. The tradition is confirmed by the curious fact that many antique Sards exhibit on the back a central circle of a different density from the rest, leading to the inference that the stone was primarily generated in a spherical shape. At present the rare sap-green Jasper is only known in its native state as occuring intermixed with fragments of others in a peculiar Egyptian breccia; out of which the old engravers must have similarly extracted their supply. Fronto seems to allude to this practice of extracting gems out of a *breccia*, when, in speaking of the judicious choice of words, he has the simile: "Verba prorsus alii vecte et malleo ut *silices* (building-stones) moliuntur, alii autem cælo et morculo ut gemmulas exsculpunt" (Ep. iv. 3). Afterwards they were discovered in many other places, such as Paros and Assos. Those from Leucas in Epirus, and the Egyptian kind, were set with a gold-foil under them; whence it may be concluded they were very clear. The Arabian were more opaque;* the Indian transparent. Of the latter there were three kinds: the red; the second either of a large size or of a fatty nature, for the passage is disputed—Jan reading "quas *Pionias* vocant a *pinguetudine*;"† the old editions *Dioneas—magnitudine;* the third was set with a silver-foil, hence must have been of a dark-red, the same foil being now in use for Carbuncles. Sards retained their lustre longer than any other gem, but suffered most from contact with oil.

Veltheim has carefully investigated the subject of the ancient trade in this and similar stones, in his treatise, 'Ueber die Onyxgebirge des Ctesias.' At that time (1797) these Sards were occasionally brought to Amsterdam and Copenhagen from Cambaya, as ballast in large casks. They had been sold at Brunswick by Voigt of Amsterdam at the rate of 38 florins per cwt. Specimens from these, in the cabinet of Dr. Brückmann, were the true Sarda Nobilis;

---

* The famous intaglio of Sextus Pompey by Agathangelus (Berlin) on a blood-red Sard was, according to Winckelmann, found to be backed with a gold-foil when drawn from the massy 'annulus unciarius' it adorned when first discovered.

† Compare the German "Speckstein," *bacon-stone*, our steatite.

and some also exhibited fine Onyx layers. He further mentions that a quarry of Sards and Calcedonies was then worked near Guzerat.* When Garcias ab Horto wrote, the Sard and the loadstone were commonly exported from Cambaya, and sold by the *manus*, or weight of 26 pounds.

Mohammed Ben Mansur divides the Carnelian into seven kinds, viz.: the liver-coloured, rose-red, yellow, white, black, blue, bicoloured (evidently including certain Agates under this denomination). He adds, " Although a hard stone, it is commonly used for the engraving of signets upon." It was then found at Senaa and Aden in Yemen; on the confines of India and Rum (*i. e.* the Byzantine Empire), and in the vicinity of Basra.

Amongst the other recipes for falsifying gems, alluded to by Pliny, one was " how the Sardonyx may be made out of a Sard." He seems here to have in view the artifice by which the former gem (and literally Pliny's own definition of it, "a white layer upon Sard," " candor in Sarda ") was imitated by placing a Sard, first coated with carbonate of soda, upon a red-hot iron. This process converted the surface into an opaque white layer of any depth required (depending upon the length of the calcination), which forms a good relief to the intaglio cut through it into the transparent ground beneath. Doubtless this effect of fire upon the Sard was discovered by accident;† and the method did not come into general use till a late period

---

* For the present state of these mines, see p. 228.

† The ancient practice of committing to his pyre his choicest ornaments with the corpse, has supplied numerous, and to collectors lamentable examples of the effect of fire and soda (derived from the wood-ashes) upon the constitution of the several gems. I have quoted Cynthia's Beryl thus sacrificed to her manes. Sards, the backs and edges of which had been protected from the action of the soda by their metal casings, are perpetually to be met with, having their surfaces to a certain depth calcined into a complete calcedony.

of the Empire, inasmuch as fine engravings upon this imitative medium are not to be met with, though mediocre works in it abound. As might be expected, it was a favourite material with the Italian artists of the Renaissance, to whom it was recommended by the lively contrast of colours it afforded when engraved. By first producing a white stratum of extreme thinness, and cutting the design out of this, as it were, in the lowest relief, the result had the appearance of a picture in opaque white upon a ruby ground, strikingly effective. The finest example of this ingenious method known to me is a Sard medallion (Marlborough), representing Christ's entry into Jerusalem. Carnelians of this date also are to be seen with mottoes and devices written upon their surface in opaque white characters very neatly done, and apparently involving incredible labour and dexterity in their performance. The recipe for this really facile operation given by Barbot, is first to obtain the white ground by the above-named process; then stopping out the parts not wished opaque with a cement containing oxide of iron, the stone on a second application of heat recovers its lost colour in those parts, whilst the letters continue opaque and white as before. The Sard has a great affinity for iron-oxide, from which its native colour is derived. It has been observed that such as have lain for centuries in their oxidised iron settings have imbibed a singular depth of tint from the contact.

Epiphanius records the prevailing belief that the Sard was of virtue for the cure of tumours, and of all wounds made by iron. Marbodus, in the 11th century (translating Evax), declares that the Carnelian drives away evil spirits, and preserves concord; whilst that of the colour of raw flesh will stanch hæmorrhages, whether natural or from wounds. As usual, this list of virtues was marvellously

swelled in the progress of the Middle Ages, so that we find
Albertus Magnus affirming that the Sard exhilarated the
soul, drove away fear, baffled witchcraft, and was an antidote to all poisons arising from the corruption of the blood;
besides its ancient property as a styptic.   Cardan asserts it
gives success in lawsuits, and makes the wearer rich.   The
philosophic De Laet, in 1647, though ridiculing all these
fables which he quotes, declares from his own experience
its power to stop the bleeding from the nose; and that
rings were cut entirely out of it to be worn for that object.
Such are still made and worn in Italy, and with the same
idea.

SARDONYX: Σαρδόνυξ: ὄνυξ Σαρδῶος.

THIS name is of later date than Theophrastus, by whom the stone was included (if indeed then known) amongst the varieties of the ὀνύχιον. It is defined by Pliny as originally signifying a white mark in a Sard (candor in Sarda) like the human nail placed upon flesh, and both of them *transparent.** Such was the true Indian sort, according to Ismenias, Demostratus, Zenothemis, and Sotacus. The two last writers call the *opaque* stone of this class "The Blind Sardonyx;" to such in Pliny's age was the name *exclusively confined* (quæ nunc abstulere nomen). The Arabian kind exhibited no traces of the Sard (*i. e.* of a *transparent* red layer), and admitted of a variety of colours; the base (radix), black or verging on blue; the surface (unguis), like vermilion, encircled by (redimitus) an opaque, fatty, white layer (intermediate), and passing into the white with a slight tinge of purple (or blue). And more tersely and intelligibly Solinus: "In the Arabian Sardonyx the surface is admired if of a purer red; but found fault with, if of a dirty tint: the middle is girt by an opaque white line. The best is where neither layer diffuses its colour upon its neighbour, nor borrows anything from the other, and the last (the base) finishes off with

* The French mineralogists still continue to distinguish this from the Sard-Onyx, by giving it the name of Sard-Agate. The upper layer (in which the designs are usually relieved) is a true, transparent, red Sard; the ground an equally transparent White Agate.

black. For any part to be transparent is reckoned a defect; complete opacity increases its beauty" (55).

Three colours were now considered essential to the idea of a Sardonyx, hence termed by Lucian ψῆφος τῶν τριχρώμων ἐρυθρὰ ἐπιπολῆς, as appears also from what Pliny says as to the manner of forging it (xxxvii. 75): "Sardonyx gems are made up out of three stones cemented together so neatly that the fraud cannot be discovered, by selecting one a black, another a white, the third a red, each one the best in its respective kind. In the same sense Martial makes his lady-killer boast of a *real* Sardonyx thrice girt with zones—

" Sardonycha verum lineisque ter cinctum,"

a gift from some fair victim. And to conclude, the Roman notion of what a perfect Sardonyx should be is prettily enunciated by Achilles Tatius, in his description of Calligone's necklace (ii. 1): "In its centre, between an Amethyst and a Hyacinth, was a triple gem, the layers of which alternated in colour, and all formed one whole. The base of the gem was black; the middle substance, white, appeared next to the black; the last, crowning the white, showed red as fire. The gem encircled within a wreath of gold, gleamed forth like a golden eye."

Though *three* layers at least were required to constitute a true *Sardonyx* (with only *two* it remained an *Onyx*), yet these might be repeated indefinitely without altering its designation. Köhler justly lays down "that it was a Sardonyx, as long as the different colours lay in regular layers one over the other. It was a Sardonyx, whether the white stratum was united with a male or female (dark or light) Sard; whether the stone possessed three, four, five, or more strata. For the name Sardonyx implied the regular union of the Sard with a white layer: now the Sard exhibited innumerable gradations, into red, yellow, brown, and black." From this explanation we can understand Pliny's observation that the base (substratum) of the Indian kind was found of a wax, or a horn colour; and the middle zones (circuli), sometimes colourless and transparent (abbi), with a certain faint iridescence; whilst the surface was "redder than a lobster's shell." Those from Armenia,* fine in other respects, had the middle zone too pale. The merit of the Arabian lay in a brilliant whiteness of this zone, which was of considerable width (non gracili), and which beamed forth (ridente) not upon the slope or low down in the gem, but upon the very projection (umbonibus); the base also being of the deepest black. Köhler, misled by the expression *umbonibus*, which he takes for the surface of the gem (which Pliny always terms *unguis*, or *superficies*), interprets this definition as if intended for our Nicoli, to which it does not by any means apply when rightly understood. Besides the fact that Pliny is speaking here of a stone necessarily having *three* layers, his *thickness* of the white zone is a sufficient refutation to Köhler's hypothesis, the top layer in the Nicolo being of infinitesimal tenuity.

Zenothemis recorded that the Sardonyx was held in no estimation by the Indians, who, instead of using it in jewelry, made sword handles out of it, so large were the pieces in which it occurred. These, it was well known, were picked up in the beds of torrents. They first came into request in Europe, when it was observed that they were almost the only class of gems which when engraved did not pull away the (soft) sealing-wax.† The natives at

---

* Whence probably the early Persians got the material in which some cylinders (though of extreme rarity) are to be found.

† Here we have the reason supplied us for the finest early engravings, whether called Etruscan or Greek, being so often found upon the Sardonyx, cut traversely: now improperly designated a Banded Agate.

(G)   U

last had learnt from the Romans to admire these stones:
" they, that is to say the lower classes,* wear them bored
and strung for necklaces, but make no other use of them.
This perforation is now considered the sure mark of the
Indian kind."

This last remark satisfactorily accounts for the fine
hole so often to be discovered passing through the axis
of Sardonyx camei. The stones imported in the shape of
oval beads were subsequently cut down into flattened
disks, so as to present their layers properly disposed for
the working out of the artist's design. From ignorance
of the original destination of the material, many antiquaries have been puzzled to explain the object of these
minute perforations; and still more the method by which
they had been drilled through such thin slabs without
splitting them. The commonness of the stone in India
appears from the exaggerated tale repeated by Philostratus (Ap. Tyan. iii. 27): " The stones that come from
India, amongst the Greeks indeed are set in rings and
necklaces, but amongst the Indians are of sufficient size
to make flaggons and wine-coolers, and bowls holding
enough to quench the thirst of four droughty men upon a
summer's day."

As the principal merit of the substance lay in the
evenness and contrast of the colours, the Romans rarely
engraved *intagli* upon it. They employed it in their
jewelry and rings, cut either *en cabochon* or in a truncated
cone of an oval section, more or less high, so proportioned
as to display the three zones to the greatest perfection;
making them all, by grinding down the surface and the
base to the due thickness, as nearly equal as possible:

* A curious proof of the permanence of all usages amongst the Hindoos, the chaplet, the *ex officio* badge of a fakir, is to this day made
of Sardonyx-beads.

it was, says Pliny, a great defect if the white zone spread
unevenly, and did not keep itself together in a perfect
band. When intagli are found on a Sardonyx, they are
always sunk but slightly into, or rather sketched upon,
the surface, so as to be invisible at a trifling distance.

It was for *camei* that this material was specially reserved;
the various shades, taken advantage of with singular dexterity by the artist, enabled him to add the charm of colour
to the relief.* This may be considered a symptom of declining taste; certain it is that such works chiefly belong to
Roman and Imperial times. The most remarkable example
that I have observed of this skilful adaptation of the design
to the material is a quadriga in the French Collection,
where each of the four horses is worked out in a perfectly
distinct colour. The Marlborough Antonia too, admirable
in point of art, exhibits a most judicious employment of
two shades, black and brown, and of a milky white, in the
hair, background, and the flesh.

The largest slab of Sardonyx known is that forming the
Carpegna Cameo (in the Vatican), 16 inches long by 12
deep. The subject, the "Triumph of Bacchus and Ceres,"
in a car drawn by Centaurs, is executed with great spirit in

---

* When Earinus, Domitian's favourite, sent his hair, a votive offering,
in a jewelled casket—" gemmata pyxide "—to his national god, Ascle-
pius of Pergamus, the gems decorating it were doubtless camei engraved
with appropriate designs. Were these groups of Cupids? for Statius
introduces them as acting as tonsors on the occasion—

> " Tunc junctis crinem incidere sagittis
> Atque auro gemmisque locant."

The casket also contained his mirror—

> " Nobile *gemmato* speculum portaverat auro."

And on its reflecting surface the boy's portrait was engraved (a photo·
graph by anticipation)—

> " Tu modo fige aciem et vultus hos usque relinque,
> Sic ait et speculum reclusit imagine rapta."

u 2

a stone of five layers; the respective parts of the design being kept in very flat relief, for the sake of contrast, owing to the thinness of these layers.

Next in size, 13 × 11 inches, but far above it in fame, comes " Le Grand Camée de France," known also as the " Agate of the Sainte Chapelle," representing the triumph of Germanicus, where the hero is received by Tiberius and Livia, and the Apotheosis of Augustus occupies the upper portion of the composition; captives seated amidst arms filling up the exergue. This was brought from Constantinople by Baldwin II., the last Frankish emperor, and pawned to St. Louis, together with other relics (of vast sanctity but no intrinsic value), for an enormous sum:* being then interpreted as " the Triumph of Joseph in Egypt." This stone also has five strata.

Third in point of magnitude, but superior to both the preceding as a work of art, in consequence of the priority of its date, is the " Gemma Augustea" of Vienna, the subject of which is the reception of Drusus (father of Germanicus) after his victory over the Rhæti and Vindelici, B.C. 17. Augustus as Jove, Livia as Roma, seated on thrones, welcome the hero and his brother Tiberius. In the exergue appear his Macedonian and Thessalian allies, with their captives; Romans are engaged in erecting a trophy over the vanquished barbarians who lie bound upon the earth below: on the shield of the trophy is blazoned the Scorpion, the horoscope of Tiberius; an ingenious mode of recording his participation in the success. This is the finest work in relief extant; the stone, however, has but two layers, and is rather to be termed an Agathe-onyx. Its shape is elliptical, 9 × 8 inches. This relic, wrested by Philippe le Bel from the Templars, and presented by him to the Abbaye

---

* 10,000 marks of silver, about 20,000*l.*, equivalent to fifteen times that sum at present.

of Poissy, was stolen thence in the civil wars of the
16th century, and ultimately purchased by Rudolf II. for
12,000 ducats (6000*l.*), and that more for its value as a
mineralogical rarity than as a work of art.*

The Marlborough Cabinet boasts of another approaching
to this in dimensions, being 8 inches wide by 6 deep, and
the most remarkable of all in a scientific point of view;
for it presents strata.of transparent sard, *purple* or rather
lilac, opaque white, and a ground of opaque black—colours
not united in any other example known to the world. It
bears in flat relief two well-executed imperial busts, in a
style marking the commencement of the Decline, attributed without much reason to Didius Julianus and Manlia
Scantilla (but more probably representing Commodus and
Marcia): the one in the combined characters of Jupiter,
Dodonæus, and Ammon; the other, of Libera, or Ceres.
One convincing reason that the prince here immortalised
is *not* Didius, is the fact that his tenure of empire only extended to sixty-five days: the execution of a cameo like
this required the labour of several months. *Guay* was two
entire years engaged upon his of Louis XV. Another
large oval Sardonyx, in the same collection, with the hel-
meted head of Constantine, is of remarkable beauty, though
inferior in size to the ægis-covered bust of Constantius II.,
the glory of the Royal Dactyliotheca, an oval of $7\frac{1}{2} \times 5\frac{1}{2}$
inches, surrounded by an architectural border : in this stone
the respective layers are of considerable thickness; in the
former they are extremely thin.

The long and luxurious reign of Constantine has left us
several other works, some of Christian character, on pieces of
Sardonyx of the finest quality; and equally beautiful speci-

* The same cabinet possesses the grand Ptolemaic Eagle *proper*, on
a circular Sardonyx, 9 inches wide. The Blacas, a bust of Augustus,
converted into a Constantine by the addition of a gold and jewelled
diadem; equally conspicuous for art and material ($6 \times 4\frac{1}{2}$).

mens appear in the latest ages of the Byzantine Empire, but disfigured by designs in the wretched taste of that period, which, from their subjects, the Angelical Salutation, the Annunciation, figures of saints, &c., were evidently intended to ornament ecclesiastical plate.

The existence of this material at Constantinople to so late a period shows what is also manifest from other sources, that a direct trade with India was maintained by the Greeks down to the times of the final extinction of their empire. Amongst the presents, the most costly in the estimation of the age (including a case of " named " relics of saints), sent by Alexius Comnenus to the Emperor Henry III., a vase of Sardonyx, and another of Crystal, held a conspicuous place. The artists of the Revival executed some camei on a scale rivalling the dimensions of the most famous antique pieces. First amongst these stands the work of *Gio. Ant. dei Rossi*, a cameo circular in form and the third of a braccio, or 7½ inches, in diameter, representing in half-length Cosimo I. and his consort Eleonora, supporting between them a shield charged with the lily of Florence, and accompanied by their five sons and two daughters.

The exact locality that supplied the Romans is indicated by Ptolemy, thus : " The most remarkable mountains in the known parts of India are the Apocopi Hills, called also the *Pains of the Gods*, and the Mount Sardonyx, in which the stone of the same name is found, its meridian being 117° 30' 21." The country between Mount Sardonyx and Bittigus is occupied by the Tabassi, a great nation." Veltheim thinks this to be the range running between Ozene (Ougein) and Tagara (Dowlatabad).*

---

* Besides paying in gold and silver, of which this trade ruinously drained the empire, the Roman merchants bartered amber, coral, copper, and lead, for these Indian commodities. " India neque æs neque plumbum habet gemmisque ac margaritis suis hæc permutat." Plin. xxxiv. 48.

The so-called Arabian kind may have been brought from
the interior of Africa by the caravans—a traffic established
from time immemorial—into Abyssinia, and the other
regions upon the Red Sea. The Ethiopian traders made
Meroe their entrepôt, where, says Diod. Siculus (i. 33),
" all sorts of precious stones were to be got ;" certainly not
found in that neighbourhood but brought by the traders.
In the times of Justinian, his allies, the Abyssinian Axu-
mites, traded from their port of Adulis all along the East
African coast as far as the equator for gold-dust, emeralds,
and aromatics. Cosmas adds that they even ventured as
far as Taprobane ; but I suspect he means Madagascar, not
Ceylon, by this vague appellation. Böttiger relates that
Sparrmann, on his return from his African exploration, gave
two pieces of the true antique Sardonyx, knocked off a rock
which had served him for a dinner-table in the desert
behind the Cape of Good Hope, to the mineralogist Danz,
who afterwards got 10 carolines for each at Vienna. Am-
mian (XIV. 3) incidentally affords us a glimpse of the
extent of this caravan trade, converging to a spot in the
heart of the Eastern empire: "*Batne*, a town in Anthe-
musia, founded by the old Macedonians, situated upon the
Euphrates, and crowded at that time (when the satrap
Nohodares attempted to surprise it) with wealthy traders ;
to which place, upon the great annual festival held at the
beginning of September, a vast number of people of all
conditions assemble at the fair, in order to purchase the
goods sent by the Indians and Chinese, and also the numerous other commodities accustomed to be conveyed thither
both by sea and land."

Veltheim, quoted by Böttiger, explained the existence of
the material in pieces of such marvellous dimensions as the
camei above described required, by supposing them formed
out of slabs of Obsidean, hardened by a boiling in sulphur,

and then polished and cemented together; not explaining, however, how the white and the red layers were produced. A sufficient answer to this hypothesis, is the circumstance that the strata in our camei often *undulate*, besides being of irregular thickness; proof demonstrative that they were arranged by the hand of Nature as we now behold them.

The Romans, however, like their descendants of our day, imitated the Sardonyx by the fusing together three pieces of coloured opaque glass, so as to represent the natural strata, with surprising accuracy (VITRUM ANNULARE). These, when worked over and polished like a gem, can only be detected by the test of the file. It is curious, however, to notice that the same idea as that propounded by Veltheim with respect to the artificial origin of the Sardonyx appears to have prevailed in the days of Theophrastus; at least this seems the natural interpretation of the passage (61): "Earthy minerals: these assume all kinds of colours by reason of the diversity of the subject-matter, and of the influences acting upon it; of which, some they soften by fire, some they fuse and pound, and so put together those stones that are brought from Asia." Now it must be remembered that both the Murrhina (porcelain Jasper) and the *gemma*, of which the huge draught-board (carried in Pompey's triumph) was made, were unknown in Rome before the conquest of Asia, nearly three centuries after the times of Theophrastus: well, therefore, might the true nature of similar substances have been a mystery to the Greeks in the reign of Alexander. The use of the Sardonyx, Pliny records, was first made fashionable at Rome by Scipio Africanus the Elder: the favourite gems of the Emperor Claudius were the Sardonyx and the Emerald. Such a predilection may well account for the existence of so many cameo portraits of this Cæsar and the members of his family in the former material. These

camei seem by a refinement of luxury to have been employed
in some manner in the decorations of state-rooms, enchased
in the furniture seems the most reasonable conjecture.
Statius, in describing the villa of Manlius Vopiscus at
Tivoli (Syl. I. 3), has—

> " Vidi artes veterumque manus variisque metalla
>   Viva modis. Labor est auri memorare figuras,
>   Aut ebur, aut dignas digitis contingere gemmas."

A perfect Oriental Sardonyx is still required to exhibit
the same characters as when Pliny defined them. The base
must be black (in reality a translucent chocolate colour
when held against the sun), the middle zone opaque fatty
white, the surface a light brown or red. It is, perhaps,
now more valuable when of a certain size (the source being
lost) than in Roman times—one as large as a crown piece
selling for 30*l.* on the Continent.* The noble collection
of Baron Roger includes two specimens of unparalleled magnitude and beauty, being circular disks three inches in
diameter, the strata perfectly even, valued at 1000*l.* each.
But that even under the Empire it was a stone of price,
however small its dimensions, appears from the provision
in Ulpian (De Bon. Damnat.), from a law of Hadrian's,
which excepts the contingency of the (capitally convicted)
criminal's having on his finger a *Sardonyx* or other precious
stone exceeding the value of 5 aurei (5*l.*).† In Juvenal's
days it was the most fashionable of all gems. Talking of
the necessity of a barrister's keeping up appearances if he

---

* Barbot quotes some fine specimens of 1¾ and 2 inches in diameter
as worth 80*l.* a piece. It must however be observed that this valuation
of the gem only holds good for the Continent. English jewellers,
lacking the skill to distinguish the precious Indian from the vulgar
artificially-stained German species, hold both sorts equally cheap.

† In such a case, instead of becoming the fee of the executioner, it
was escheated to the fiscus.

wished for clients at Rome, he adds: "Put not trust in
your eloquence; no one nowadays would fee Cicero himself with a couple of gold-pieces unless a big ring glittered
on his finger, and for this reason Paulus used to *hire a
Sardonyx* to plead in, and on the strength of it got higher
retainers than Cossus or than Basilas." A *real* Sardonyx
was likewise the most acceptable mark of gratitude, like a
diamond-ring nowadays, that could be offered to a successful
barrister by a fair client (Mart. x. 87):—

> "Infamata virum puella vicit?
> Veros Sardonychas sed *ipsa* tradat—
> Mirator veterum senex avorum
> Ducat Phidiaci toreuma cæli."

The finest specimen, however, of the kind is the "great
Sardonyx" of the Green Vaults, Dresden, an oval, 6 × 4½
inches (intact) mounted as a pendant, with three smaller
but still magnificent fellows of the same quality hanging
from it.

In the Temple of Concord there was shown to Pliny's
contemporaries the supposed ("si credimus," as he qualifies
the *custode's* story) gem of the celebrated ring of Polycrates
—a Sardonyx, not engraved (intacta, illibata), set in a
golden cornucopia (held by the goddess), the gift of
Augustus, where it occupied almost the last place amongst
a crowd of other gems, all deemed of superior value. It is
difficult to explain how this legend came to be affixed to
this particular Sardonyx; for Herodotus expressly calls the
signet in question "an *Emerald*, the work of Theodorus the
Samian," Clemens Alexandrinus adding that the device
upon it was a lyre. And Pausanias has (viii. 4): "A work
of Theodorus was also that signet upon an *Emerald* which
the tyrant Polycrates chiefly wore, and upon which he
prided himself excessively." All testimonies showing that
the *Emerald*, priceless both from its intrinsic worth and the

engraving upon it by so famous an artist, was the object
the sacrifice of which to Nemesis was deemed sufficient to
atone for all the other over-great favours of Fortune, and to
avert her jealousy.*

The Sardonyx is said by Marbodus to typify three of the
Cardinal Virtues: the black, Humility; the white, Chastity;
and the red, Modesty or Martyrdom. Strange to say, however, in spite of its beauty and high estimation, it possessed
no medicinal or mystic qualities.

(Köhler: 'Untersuchung über den Sard, den Onyx, und
den Sardonyx.' Brückmann's reply to him in his 'Nachtrag
über den Sarder, &c.' Böttiger: 'Ueber die Echtheit und
das Vaterland der Antiken Onyxcameen von ausserordent-
licher Grösse.' [A most copious and instructive dissertation.]
Veltheim: 'Etwas über die Onyxgebirge des Ctesias.')

* Hardly a relic of the kind equals in historic, nay, romantic interest,
the fine, simply-polished Sardonyx (2 inches wide) discovered by Mr.
Newton in a drain under the Mausoleum, into which it had doubtless
been tossed by the mediæval despoilers of the royal corpse, after stripping
the gem of its gold framework. It had evidently been worn as Calligone's above quoted.

SOLIS GEMMA: *Adularia.*

THE definition of the " Solis Gemma," given by Pliny (67) as " colourless, but diffusing brilliant rays in a circle, after the fashion of that luminary," exactly describes the Adularian Felspar, now, however, known to lapidaries as the Moon-stone, from the silvery radiance of the large orb that illumines its convex surface. Sometimes the ground has a slight tinge of the pale blue peculiar to the asteria Sapphire. Though now abundantly produced in Mount Adula (St. Gothard) of the Alps, whence its epithet Adularian, the best examples come from Ceylon : therefore there can be little doubt the Romans had received consignments of it amongst the other products of Taprobane. A very curious variety, the " Fish's Eye," greenish in tint, becomes red by transmitted light, going through the same changes as a bead of schmetze paste.

Pliny's Selenites (Moon-stone) seems to be a variety of the last, " shining with a yellow lustre from a colourless ground, containing an image of the moon, which (if the story be true) daily waxes or wanes, according to the state of that luminary.* It is said to be found in Arabia." And

* The original notion doubtless being that it was the Moon's *image* in the gem that sympathised with her changes, the mediæval love of the marvellous readily transferred this property to the entire bulk of the stone. It must be observed that Marbodus describes its colour as that of the Jasper or the Emerald—another indication that he had the true Ceylonese gem in view.

Psellus also, much to the same effect: "It is so named because it displays as it were an eye within itself, which increases or diminishes according to the growth or decline of the moon." Marbodus, improving upon this, makes the *stone itself* increase or decrease proportionally with the moon, whence it derived the title of the Sacred Stone.

Orpheus, however, differs from Pliny in his description of the "Gem of the Sun" (388), of which he makes two kinds. " In both, there grow real rays, straight, shining, and in appearance like unto hairs ; but the colours of the stones are different : one of them you would deem a Crystal, but the other is exactly like the Chrysolite, and if it did not contain the hairs it would be a Chrysolite." Now, the first of these is evidently the Venus-hair Crystal—a Crystal of Quartz of extreme brilliancy, full of long straight fibres of asbestos, like the finest hairs inclosed within its sub-stance. This attribution is confirmed by an intaglio dis-covered by me amongst the Marlborough Gems—a figure of Sol engraved upon such a Crystal, marking it for that god's own peculiar property. The yellow variety it is more difficult to identify.

The modern Venus-hair Crystal is not, however, the Veneris Crinis of Pliny; for that was " extremely black and lustrous, containing in its substance the appearance of red hairs," characters assigning it to the Jasper species.

No wonder that so singular a phenomenon should have been regarded as the sign of wondrous inherent virtues. " Phœbus hath planted in both species a mighty spirit that gives glory and heroic beauty to whosoever shall wear them with due understanding," adds Orpheus.

The " Sun-stone " of our jewellers only differs from their " Moon-stone " in colour, being of a very lustrous brownish-yellow, instead of a faint bluish-white. In the opinion, however, of a very experienced lapidary it is not a felspar at all, being considerably harder, and to all appearance a variety of the yellow quartz. There can be little doubt that this was Pliny's *Selenites*, but its yellow splendour has occasioned its transference from the pale Dian to the golden Sol. The Selenites, with the same idea of lunar sympathy attached to it, ranks amongst the earliest of stones recognised as precious, for it adorns that most ancient of all jewels, the necklace of Harmonia ('Precious Stones,' p. 306). But the later Greeks give the name (and *Aphroselenon*) to a totally different thing, *Talc* (which we have borrowed from the Arabic), because "only to be discovered by moonlight;" though the resemblance of its sheen to that luminary's were a better and a more probable derivation. It singularly exemplifies the persistence of old religious beliefs under altered forms to find the Germans still calling Talc, "Marien-Glas," and "unser-lieben-Frauen-Eis," "St. Mary's glass, and "Our Lady's ice."

SUCCINUM : Ἤλεκτρον : *Amber.*

AMBER claims the highest antiquity in the list of Precious Stones used for personal ornament. It was the only one known to the early Greeks. Homer mentions no gem in his minute description of various jewels, except this, " the gold necklace hung with bits of amber " (μετὰ δ'ἠλέκτροισιν ἔερτο), offered by the Phœnician trader to the Queen of Syra (Odys. xv. 460). It is seen, often carved into elegant forms, in the most ancient Etruscan jewelry. Amber scarabei alternate with others in Sardonyx, as pendants to the magnificent necklace known as the Prince di Canino's, the masterpiece of the Etruscan goldsmith. This people, becoming acquainted with the substance from their inter- course with the Gauls, had speedily availed themselves of it as a glyptic material, charmed by its extraordinary lustre, so far superior to that of the ill-polished gems that lay within their reach. Even in the best days of Greece it was held in the highest estimation. Theophrastus (29) speaks of " the Fossil Amber, a stone produced in Liguria, possessing , the property of attraction, but rare, and found in only a few countries." Even in Pliny's age, he satirically observes (xxxvii. 12) that " the price of a figurine in it, however small, exceeded that of a living, healthy slave," which we know was by no means inconsiderable. Juvenal represents his rich patron as displaying at his feast a bowl embossed with Beryls, and relievi in Amber " Heliadum crustas." Heliodorus makes the brooch fastening the mantle of his hero Theagenes "a Pallas carved out of amber;" the bust of the goddess it may be presumed. The most precious example extant of Roman carving in this substance is a ring (Waterton) formed into an elegant design with Cupids in full relief upon the shoulders, cut out of a single piece.

To the very last, as we see from the article in Epiphanius on the "Lyncurium," there is always an uncertainty whether the Greeks understood Amber or Zircon under that name, or its other form Langurium. Though Theophrastus (28) sufficiently shows that his Lyncurium is a Zircon, by mentioning its extreme hardness, and employment for signet-stones, yet Pliny (13) asserts that, if the Greek mineralogists did not understand Amber by that name (otherwise, *Langurium*), there was no such thing in existence ; adding that the whole account of its origin was a mere fable.

Pliny, after craving the indulgence of his readers, takes occasion to expose "the love of lying distinguishing the Greeks, since there was some advantage in knowing all their marvellous tales;" and proceeds to enumerate the various fictions by which they had accounted for the production of Amber. The most ancient of these, resting chiefly on the authority of the poets, made it to be the tears which the Electrides, changed into poplars along the banks of the Eridanus, dropped upon the death of their brother Phaëthon. Others made it distil from trees in certain islands far up in the Adriatic. The explanation of its being the condensed urine of the lynx had also many supporters, and the habitat of these beasts was placed in the north of Italy.* Sudines and Metrodorus endeavoured

---

* The North of Italy was the same Terra Fabulosa to the earlier Greeks, that India became to the later, as the knowledge of geography extended.

to escape the absurdity of this hypothesis by sagaciously
explaining that the tree producing it in Liguria was called
the *Lynx.* ·Numerous other accounts follow : some placing
the mines in Scythia (the favourite ancient cloak for
ignorance); others in India; some, lastly, in Egypt, under
the name of *Sacal,* a notion where seems to lurk a confusion
between Amber and certain odoriferous resins.

Pytheas, however (about 280 B.C.), finally brought back
from his voyage the correct account that the Gutones, a
German tribe dwelling upon the estuary Metonomus, picked
it up in spring upon the coast of the island Abalus, or
Basilea, a day's sail out at sea, and either used it for fuel
or sold it to the Teutons. This explains the prevalent
Greek notion of Liguria being its native country, inasmuch
as from the Northern Sea (the Baltic, the sole locality producing it in marketable quantities) it was carried overland
by the Germans into Pannonia, and thence diffused amongst
the Heneti, a nation seated at the head of the Adriatic.
From these it was procured by the Cisalpine Gauls, a
kindred race; whence originated the fable of its being
carried down with the stream of the Po. In Pliny's age it
was universally worn in necklaces by the Transpadane
females (of Lombardy and Piedmont) partly as an ornament, partly as a prophylactic against *goîtres,* to which
they were subject in consequence of the bad quality of the
water. Thus the early Greeks obtained it from the Teutonic tribes upon the Adriatic; the Etruscans from the
Cisalpine Gauls upon their northern frontier.

It is strange the material should have maintained its
high value amongst the Romans, in spite of the enormous
importation that had gone on ever since the German cam-
paign of Germanicus had opened up a communication with
the Baltic. That commander had actually visited with
his fleet the island Austeravia, that furnished the chief

(G)   X.

supply. This place the Romans called *Glæsaria*, from *Glæsum* (whence our " glass "), the native name for Amber. This locality was distant 600 miles from Carnuntum in Pannonia (Altemburg on the Danube), apparently mentioned here as being the ancient entrepôt where the Teutonic and Roman traders met. When Pliny wrote, the Roman knight was still living who had been despatched by Julian, superintendent of Nero's gladiators, to investigate that coast, and the nature of the trade there. So successful did his expedition prove (Solinus stating that the king of the country sent Nero a present of 13,000 pounds weight of the precious commodity) that all the weapons and articles used during the games of the amphitheatre, on one particular day, were made of Amber, and the network protecting the lower tier of seats was knotted with it : that is, the cords forming it were passed through perforated Amber balls at their intersections. The largest mass imported on this occasion weighed 13 pounds. The greatest weight on record is 22 pounds, being that of a lump found a few years ago between Memel and Königsberg, on the Baltic. This district, as in Nero's age, is the sole one that furnishes Amber in marketable quantities (to the amount of 4000 pounds' weight yearly), although small specimens are found in all parts of the world from China to Hampstead ; in fact, wherever beds of lignite, its true matrix, occur.

Tacitus (Germ. 45) gives the name of the Amber-gatherers as the Æstyi, with the remark that they are a sacred nation, worshippers of the mother of the Gods (Hertha), and resemble the Suevi in customs and appearance, but speak a language like the British. As the symbol of their faith, they wore the figure of a wild boar, protected by which they could travel everywhere in safety, even in time of war.

Pliny mentions a singular employment of this substance,

which was in the imitation of all the transparent precious stones, but above all, of the Amethyst; for it could be stained of any colour required. This was done by boiling it in kid's fat and alkanet-root, or else in the murex-blood. Before the wear of its tender surface betrayed the fraud, it must, thus prepared, have passed for a precious stone of unusual brilliancy, for in refractive power it is only second to the Diamond. In fact, so great is the analogy between both, that a theory has been started, and endorsed by eminent scientific names, that the Diamond also is no more than a fossil resin. This idea is supported by the late discovery in such abundance of its black cogener, the Brazilian Carbonado, which stands in the same relation to the Diamond as Jet does to Amber. ('Precious Stones,' p. 115.)

As in our own times the chief part of the Amber furnished by the Baltic finds its way to China, to be there used for burning in powder as incense (very little being now worked up for Europe into ornaments, no longer in fashion here), so Nicias described the Indians as using it for incense in preference to any other perfume. The Romans distinguished all the varieties we now find. The pale was believed to emit the most agreeable odour (when rubbed), but this and the wax-coloured were the cheapest. More valuable was the deep yellow (fulvus), especially if transparent and full of fire; but the best was that called Falernian, from its similarity in colour to that wine (like golden sherry), transparent with a soft lustre.

"Amongst the other monstrosities of his life," to use Pliny's words, "Nero adopted this name in his verses to express the colour of his beloved Poppæa's hair, calling it Amber"—no such very monstrous crime, one might think. From the simile we may discover that this famous beauty was a blonde with auburn hair, the colour ever esteemed (on account of its rarity amongst them) the most lovely

by the nations of the South. Hair of this colour therefore
became all the fashion amongst the Roman ladies, and false
fronts were largely imported, obtained, and doubtless without much ceremony, from the female savages of Germany.
They also changed their black locks into red by steeping
them in a powerful alkali, thus satirized by Martial :—

> " Caustica Teutonicos accendit spuma capillos :
> Captiva poteris cultior esse coma."

> " From caustic lather flames Batavia's hair :
> With captive locks thou may'st seem doubly fair."

The colour was equally admired in both sexes : Euripides
makes his Electra sneer at Menelaus, vain of the auburn
curls floating down his shoulders ; and six centuries later
it embellishes the pretty portrait Lampridius draws of the
Cæsar Diadumenian, that embodiment of the Roman idea
of beauty: " Puer fuit omnium speciosissimus statura
longiuscula, crine flavo, nigris oculis, naso deducto, ad
omnem decorem mento composito, ore ad oscula parato,
fortis naturaliter, exercitio delicatior."

Pieces of Amber containing insects speedily attracted
the notice of the Romans, and guided them, for once, to a
correct theory as to its origin ; for it was evident that it
must have enveloped these foreign bodies whilst yet a
liquid exudation from the tree. Hence they gave it the
expressive designation of *Succinum,* or the Gum-stone. The
Greek *Electrum* is referred to its supposed connexion with
the Sun-god, one of whose titles was *Elector* (" The Awa-
kener "). Martial has three ingenious epigrams upon a
viper (some small reptile), upon an ant, and upon a bee
thus imprisoned :—

> " Flentibus Heliadum ramis dum vipera repit
> Fluxit in obstantem succina gutta feram ;
> Quæ dum miratur pingui se rore teneri
> Concreto riguit vincta repente gelu.

Ne tibi regali placeas, Cleopatra, sepulcro
Vipera si tumulo nobiliore jacet."—(iv. 59.)

" Dum Phaëthontea formica vagatur in umbra
Implicuit tenuem succina gutta feram ;
Sic modo quæ fuerat vita contempta manente
Funeribus facta est nunc pretiosa suis."—(vi. 15.)

" Et latet et lucet Phaëthontide condita gutta
Ut videatur apis nectare clausa suo,
Dignum tantorum pretium tulit illa laborum
Credibile est ipsam sic voluisse mori."—(iv. 32.)

Gesner figures as the frontispiece to his book, ' De Natura Fossilium,' a ring carved out of one piece of insectiferous Amber, so ingeniously managed that the insect (a beetle) contained therein, forms the centre and ornament of the shield, as if enchased under a crystal. Mineralogists have at last been forced to revert to the theory of the first discoverers and to allow that Amber is the fossilised resin of an extinct *pinus*, the debris of which compose the lignite beds, its present locale. And chemistry has even mimicked successfully the operations of nature, and produced a true amber by boiling together by a graduated heat equal parts of rectified spirits of asphatium and of turpentine, until the compound becomes inspissated.

Its medical virtues amongst the Romans were not much more extensive than those it is at present popularly believed to possess. It was worn as an amulet by children, but Callistratus laid down that it was of service in every period of life against insanity or stranguries, either taken internally in powder or worn round the neck. This physician gave the name of Chryselectrum to the sort of a clear golden colour, supposed to be of so fiery a quality as to ignite if even brought near a flame. This kind, worn round the neck, cured the ague ; * ground up with honey and

---

* That the wearing an amber-necklace will keep off the attacks of erysipelas in a person subject to them, has been proved by repeated

rose-oil was a specific for deafness, and with Attic honey for dimness of sight.

The large circumference of some antique bowls of Amber, occasionally discovered, seems to preclude the possibility of their being hollowed out of a single block. It has been ascertained that by boiling pieces of Amber in turpentine they can be so softened as to be kneaded together into a coherent mass, or moulded to any form. Perhaps the malleable glass mentioned by Pliny (xxxvi. 68) and by Petronius ('Trimalchio's Supper') as a secret discovered in the reign of Tiberius, was some such plastic preparation of Amber made up into vases. Amber was a favourite material for the artists of the Revival to build up therewith their fairy palaces, designed for jewel-caskets, in which the Florence Galleria is extremely rich. But the master-piece in this line is the French cup, in form a long oval, pinched in towards the middle, 14 inches long, by 7 high, and valued in 1791 at 3000 francs (120*l.*).

Electrum also denoted a particular alloy of gold. " Wherever," says Pliny (xxxiii. 23), " the native gold contains one-fifth of silver, it is called Electrum. This occurs in grains in the gold-washings (Canaliense): it is also made artificially by the addition of silver; but if the proportion of the latter exceed one-fifth, the alloy does not stand the hammer. Electrum was anciently held in esteem, as Homer testifies, who makes the palace of Menelaus resplendent with gold, electrum, silver, and ivory (Od. iv. 73)." But ἤλεκτρον in this passage of Homer more probably signifies amber, as it certainly does in xv. 460. Sophocles, however, employs the word in the later sense (Antig. 1037),

experiment beyond all possibility of doubt. Its action here cannot be explained; but its efficacy as a defence of the throat against chills is evidently due to its extreme warmth when in contact with the skin, and the circle of electricity so maintained; which latter, indeed, may account for its remedial agency in the first case. '

coupling the *electrum* from Sardis with the gold of India.
By a singular coincidence, the primitive gold coinage of
the Lydians is in electrum. It would seem they knew not
the art (a difficult process) to separate the silver from the
native gold of the Pactolus washings. " Lindus in Rhodes
possesses a temple of Minerva, in which Helen has dedicated a cup made of electrum, tradition says, upon the
model of one of her own breasts. It is the peculiarity of
electrum (the metal) to be more lustrous than silver by
lamplight. The native kind also detects the presence of
poison; for then an appearance like the rainbow flies to
and fro in the vessel, attended by the crackling of flame,
and gives warning by this double indication."

TOPAZIUS : Τοπάζιος : *Peridot : Chrysolite.*

THIS gem derived its name from the island in the Red Sea,
300 stadia (37½ miles) off the mainland, where it was first
discovered ; Juba, quoted by Pliny (32), oddly enough
deriving the name from *Topazein,* which, he says, in the
" Troglodyte tongue " * means to seek after, because the
island is often lost amidst thick fogs.   According to
Archelaus, Cytis was the name of the same island, " where
certain Troglodyte pirates, when hard pressed by famine,
as they tore up the herbs and roots there growing for their
sustenance, accidentally discovered this gem." Epiphanius
has a long and confused story to much the same purpose,
but makes Topaze an Indian town, " where the stone was
found accidentally by some quarrymen, who mistook it for
Alabaster, and sold it to Theban traders, who bringing it
home, their queen set it in her diadem, upon the middle of
her forehead."

Diodorus Siculus (iii. 39) gives a detailed account of this
island.   It lay out at sea, some distance to the south of the
harbour of Aphrodite, was 80 stadia long (10 miles), and
called the Isle of Serpents, from the multitude of reptiles
formerly infesting it.   These had been completely extir-
pated by the care of the Alexandrian kings.   The Topazion
here found was a transparent gem, agreeable in aspect, re- .

---

* The Numidian king, despite his extensive learning, could never
have read the Attic tragedians, or else he would have remembered that
τοπάζειν perpetually occurs in them with this sense.

sembling glass, and presenting a wonderful golden appearance. No one was suffered to land there under pain of
death, and no boat was allowed to be kept on the island.
Provisions for the few soldiers on guard there were brought
at intervals from the continent. The gem was not discernible by day, its lustre being then overpowered by the
sun's rays, but at night made itself conspicuous by its
brightness; the guards, who divided the island among
their patrols, then ran up and covered the luminous spot
with a vase of equal size. Next day they went their rounds,
and cut out the patch of rock thus indicated, and delivered
it to the proper persons to be polished.

This stone was indubitably our Chrysolite (Peridot);
the distinctive characters of which exactly agree with those
pointed out by Pliny. His Topazius was imported from
some place in the Red Sea; at present the Peridot comes
from the Levant, but from an unknown source: it was of a
bright yellowish-green, a colour peculiar to itself (in suo
virens genere), and was the softest of all the precious
stones, yielding to the file, and suffering abrasion from
wear. Juba adds that it was found in masses of such
magnitude as to serve for a statue, four cubits high, of
Arsinoe, queen of Ptolemy Philadelphus, standing in the
" Golden Temple : " an exaggerated story, like those noticed
under " Emerald," relating most probably to some imitation in glass. It had first been introduced into Egypt in
the time of her mother, Berenice I., by Philemon " the
admiral," whom we may hence infer had discovered the
mine during his expedition against the pirates above mentioned. It was still highly valued in Pliny's age, though
somewhat fallen in estimation since the time of its first
discovery, when it was " wonderfully admired " (mire
placuisse), and preferred to all other gems.

Under the Romans it had recently been met with in the

neighbourhood of the Egyptian Thebes; and the lapidaries accurately discriminated the two varieties the Chrysopteron, our Chrysolite; and the Prasoides, our Peridot; the latter " aiming at the exact imitation of the colour of the leek-leaf." For, although chemically the same, both being Silicates of Magnesia coloured by Protoxide of Iron, yet, from the jeweller's point of view, there is a great difference between the Chrysolite and the Peridot. The former is somewhat harder, and the yellow in it greatly predominates over the green: it possesses much of the Diamond's lustre, which it exactly resembles by candle-light, when that greenish-yellow tinge is no longer discernible. The *Chrysoberyl* of the mineralogist, an extremely hard combination of alumina with small proportions of glucina and silica (whence its superiority in hardness to the other Beryls) is universally known in the trade by no other name than "oriental" Chrysolite. Science, again, restricts the designation of " Peridot " to the softer silicate of magnesia. In the Peridot green is the predominant colour, but slightly modified by yellow; in fact, in the rough it much resembles a rolled pebble of bottle-glass, or a " Brighton Emerald." * No wonder that the gem so greatly charmed the Greeks, whose sole method of cutting coloured stones was *en cabochon;* few others approach it in lustre and richness of hue, and but for its extreme softness it would still hold a high place in the rank of precious stones.†

This substance is polished with great difficulty, and only by the use of tripoli moistened with vitriol upon a leaden wheel, a secret discovered in Europe not before the last

* Hence styled by Orpheus (277) ὑαλοειδέες τόπαζοι : the common glass of the ancients having always a green or blue tinge. •

† The French indeed have a proverb, " Qui a deux Péridots on a un de trop :" but their taste here, as in the matter of the rose-coloured Pearl, has been perverted by fashion. In Italy the stone retains its ancient estimation.

century. This makes it not improbable that Theophrastus may have had the Peridot in view when speaking (27) of the peculiar process required to give lustre to the Smaragdus, naturally a dull stone ; for in his age the coast of the Red Sea was the only locality supplying the Greeks with the true Emerald, and very possibly with the Peridot, at a somewhat earlier period than is allowed by Juba. Certainly the natural brilliancy of the true Emerald prism is its most distinguishing feature, and could not have been much improved by the simple cutting of the Greek lapidary. Some fine Greek intagli occur in Peridot, to be ascribed from their style to the date of its first introduction at the Alexandrian court, but they are of the highest rarity. The Romans appear never to have used the Topazius for engraving on, deterred either by its softness, entailing the speedy destruction of the intaglio, or else by its high value as a precious stone. Modern works in it, on the contrary, are abundant enough, and to this class will the majority of supposed antiques in Peridot, when critically examined, be found to belong.

Much light is thrown upon the meaning of Theophrastus as above quoted by the following instructive chapter of Ben Mansur's :—" Abunnasser Farabi, and many other philosophers, hold the Chrysolite for no distinct species, but for a sort of Emerald, than which it is more agreeable and purer (*i. e.* more free from flaws). It is divided into three classes : the strong green, middling green, and weak green. It comes from the same mines as the Emerald, and appears to be formed of the same substance, but less perfected. Teifashi says that in his time no Chrysolites were found any longer : those now seen come from Mauritania, and are the relics of the army of Alexander. When he with his host had penetrated into the Land of Darkness, where flows the Green Fountain of Life, a voice was heard that the gravel under their feet, green from the reflexion of the fountain, was the 'Gravel of Repentance' (Hassbaen. Medamet). When they returned to the light they found this voice verified; for both they that had picked up this gravel repented, and they that had picked up none: the former, because they had got nothing but Chrysolites (instead of Emeralds); the latter, because they had got nothing at all. Hence is it called the 'Stone of Repentance.' "

Epiphanius relates a curious notion as to his Topazius, which he strangely makes a gem "red of colour beyond the Carbuncle," namely, "that if rubbed down upon a physician's hone it does not give out a red juice in accordance with its own colour, but one like milk, and fills as many cups as he that rubs it down may choose, and yet loses nothing at all of its original weight. This juice is a remedy for all diseases of the eyes, and, taken internally, for the dropsy; and is an antidote against the poison of the ' marine grape.' " *

The modern Topaz is a totally distinct substance from the ancient Topazius; being a combination of alumina and silica with fluoric acid, extremely hard (reaching 8 on the scale, while the latter barely attains to 6), highly electric, and of a vinous orange colour, without any admixture of green. Those used in jewelry come from Brazil. An inferior kind is found in Saxony. It is needless to add that it was totally unknown to the ancients, or to the mediæval writers even as late as De Boot, whose jeweller's

---

* Orpheus says that the Magi ascribe the same powers to the Topazius as to the Lychnis, viz., if thrown into a pot upon the *fire*, to prevent its boiling however hot the blaze; but to make it boil over when set upon cold ashes. Now the latter was the *red* Spinel.

Marbodus avers that the Chrysolite strung on a hair out of an ass's tail is of mighty virtue to scare away evil spirits.

Topaz is divided into two species, the Oriental and the
European,*—the Oriental yellow Corundum, and the large
Bohemian yellow Crystal.

As early as the eleventh century we find Marbodus describing under " Chrysolite," the stone now known by that
name, Pliny's Chrysopteron. He also divides the Topazius
into two kinds : one bright yellow and hard (our Chrysolite); the other greenish and soft. Thus the name
"Topaz" came to be applied to any bright yellow and
hard stone that did not to the eye display the distinctive
mark of the Chrysolite, a tinge of green, and gradually
became restricted to the sense it now bears.

The name *Peridot* can be traced far back. In the
Wardrobe Book of 27 Edward I., is entered amongst the
jewels of the deceased Bishop of Bath and Wells (escheated
to the Crown), " Unus annulus auri cum *pereditis.*" This
term implied a distinction from the Topazius; for the preceding item is, "Unus annulus auri cum topacio." The
origin of the word can only be conjectured.

An older form of Peridot is *Perithe* and Peridonius,
defined by Albertus Magnus as a gem of a yellow colour
good for the gout; and one sort of it resembling the Chrysolite. He derives the name from *Pyrites*, but without
good grounds.†

That very elegant gem the Pink Topaz of our jewellers,
emulating the Balais in tint and lustre, is not a natural
variety, but merely the dark-orange Brazilian metamorphosed by the action of fire. The change is simply and

* This was the reason why the Brazilian Topaz, when it came to be
largely imported into Europe at the end of the next century, being a
fashionable stone, and of considerable value, at once usurped the name
of a species with which it had no affinity whatever except in colour.

† Others give for the root the Arabic of the same sound, signifying
" a gem." " Topaz " has been derived from the Semitic *patazd*, trans-
posed, coming from the Sanscrit *pita*, yellow.

easily effected by closely enveloping the stone ready cut
and polished in German tinder, firmly bound with fine
wire; this being kindled, when thoroughly burnt out the
colour of the Topaz will be found converted to a clear rose
without the least detriment to its original texture or polish.
But if the roasting be carried too far it extracts all colour,
and the stone assumes the whiteness and lustre of the Novas
Minas sort, the so-called " Slave's Diamond." An extremely
rare variety of the Brazilian Topaz, shows an indigo blue,
but with a peculiar tint that ever at first sight distinguishes
it from the Sapphire. The finest specimen in existence
belonged to Mr. Hope, weighing 1¼ ounce, and was brought
by Mawe from Brazil: previously it was unknown in
Europe. The mine had only been discovered during
Mawe's residence there. The Siberian blue is paler. The
White Topaz, better known as the "Novas Minas," is perfectly colourless and very brilliant. It greatly resembles
the Diamond itself in every point except its iridescence,
and has not unfrequently been substituted by fraud for its
precious companion in the mines.

Gabelchover, in his commentary upon Baccius, mentions
(sub Topazio) that Hadrianus Guilelmus of Naples possessed one engraved in antique Roman letters with the
words NATVRA DEFICIT . FORTVNA MVTATVR . DEVS OMNIA CERNIT :
which is the only example I have met with of a spell couched
in that intelligible language. In the Blacas Cabinet is a
good head of the poet Horace on a stone described as Topaz
in the catalogue, but more probably a yellow Beryl.

ZMILAMPIS: ZMILACES: *Cat's-eye?*

DESCRIBED by Pliny as found in the Euphrates, resembling
Proconnesian marble, with a greenish-yellow colour in the
centre—terms too vague to be of much assistance in identi-
fying the gem intended, were they not elucidated by the
fuller details supplied by the more technical Solinus (xx);
" Zmilaces, a gem picked up in the bed of the Euphrates,
in appearance like the Proconnesian marble,* except that
in the middle convexity of the stone a green thing shines
through like the pupil of the eye."
    This exactly applies to our Cat's-eye, a transparent Quartz
full of minute fibres of Asbestos, and of a yellow hue
slightly tinged with green. Opalescent from its constitution, it is always cut in a highly convex form, and of a long
·oval. The opalescence is thus confined to a narrow vertical
streak of light, exactly resembling, from its contrast with
the yellowish green of the ground, the pupil of the feline
eye at noon-day. That the ancients knew this stone there
can be little doubt; for it is very accurately described,
and under its present name of Cat's-eye (Ainol. Hurr), by
Ben Mansur, whose work faithfully represents a state of
mineralogical knowledge that had existed long before his
period. He places it sixth on his scale of value, next to
the Diamond: for strange to say he omits the Opal entirely·
    Ceylon exclusively furnished it of old, as at present.

* The ' Bianco e nero antico,' according to Corsi.

Ben Mansur has : " It is asserted that the Cat's-eye is found in the Jacut-mines, and is formed out of the same substance." The latter were in the island of Saharan, lying 40 parasangs behind the island of Ceylon: and the only source of the gem (except a lately reported one near Cairo) then known. A perfect gem is yet of considerable value in Europe, but the Hindoos admire it above all precious stones after the Diamond. Even higher was its estimation with the Arabians of the Caliphate ; Teifashi remarking that of all precious stones none fetched so high a price as the Cat's-eye ; adding that one Ismael Salamita paid 700 " deniers of the elephant " for such a gem, and disposed of the same for double that amount to the Prince of Yeman.

The Hindoo love for this more curious than beautiful gem sprung from the belief in a certain inestimable influence inherent in its nature, that of never allowing the wealth of its owner to decline, but on the contrary, of stimulating it to perpetual increase. No wonder then that Garcias should note that a Cat's-eye valued in Portugal at 90 gold-pieces was sold to his knowledge in India for 600. In Europe at the time its current value was the same as that of the Opal : now far below.*

Amongst the Marlborough Gems one of the most curious is a singular conversion of a monster Cat's-eye, 1½ inch high, into a lion's head admirably carved out in full relief. The play of colours imparts to the grim mask a vivid reality of life and fury, rendering it a most successful achievement of the school (the Cinque-cento) that produced it, whose taste, ever stimulated by the love of the grotesque, revelled in similar analogies between the subject and the

---

* This old common Cat's-eye must not be confounded with the Chrysoberyl-Cat's-eye : which is the Cymophane, "floating light " (p. 38), cut so as to contract its iris from a circle into a long oval, and which equals the Sapphire in value.

material. To the same age and love for glyptic double-
entendres is due the Hope "Mexican Sun-Opal," 1 × ¾
inch in size, richly lustred with shades of red, green, and
blue, and as appropriately converted into a head of Phœbus.

The Italian name for our Cat's-eye is Belocchio, evidently
a corruption of the Beli Oculus of the Romans; but this
descriptive epithet has been transferred to the Cat's-eye on
no sufficient grounds, the Beli Oculus (Baal's-eye) having
been merely some brightly-shaded variety of the Eye-onyx:
(55) "for in it a transparent white belt encloses a black
pupil, having a golden colour shining out from the centre;
—on account of its appearance consecrated to the supreme
god of Assyria." These terms make it certain that *three*
distinct colours were necessary to compose the Beli Oculus,
whereas the Cat's-eye presents one uniform tint, the pupil
being formed by the mere reflexion of light, and shifting
about according to the angle of inclination of the surface.

The Beli Oculus of De Boot's age still retains its ancient
meaning. He describes it as a minute Eye-onyx of sovereign virtue for all diseases of the eyes (on the strength of
its name, no doubt), the mode of employing it being to
place the gem under the lid, and allow it to work its way
into the corner.

Pliny furthermore distinguishes, by significant Greek
epithets, several varieties of the Eye-onyx, as the Leucoph-
thalmus, of a fiery red enclosing the figure of an eye in
opaque white and black (for the pupil): and the same
stone seems again designated as the Lycophthalmus (Wolf's-
eye), but presenting four distinct colours, a fiery red enclosing one the colour of blood, in the centre of which
was a black spot surrounded by a circle of opaque white,
precisely representing the same organ in the wolf. The
Triophthalmus, produced with the Onyx, presented the
figure of a triple human eye.

(G)   Y

Examples of the Eye-onyx, exactly agreeing with the above descriptions, are often to be seen mounted in antique rings of the Lower Empire, and yet more so in those of mediæval times.* Having their natural form but rarely interfered with by the introduction of an intaglio, it may be concluded that they were reserved intact as amulets preservative of the wearer's sight. The most gigantic memento of Roman extravagance in the article of finger-rings, the fantastically-shaped *lion* of the Waterton Dactyliotheca ('found in Hungary) of the prodigious weight of over two ounces, exhibits for its gem a lovely Eye-onyx, shaded with blue and white. To leave no doubt of the sense in which the old jeweller had understood this work of Nature's painting, he has taken care to fashion the beasil containing it into the exact shape of the human eye.

* I observed amongst the Royal Camei a beautiful example (with two *eyes*) in a Gothic setting for a pendant jewel, its costly *enchassure* attesting the value placed upon the stone at the time. In the Blacas, two similar eyes on a rich sard ground have been cleverly transformed by a Cæsarian artist into votive clypei, containing heads of some youthful prince and his bride, and displayed by the hands of a soaring Victory.

VITRUM ANNULARE : Λίθος χυτὴ : λιθινὸν χυτὸν :
Pastes.

" PASTE," from the Italian *pasta*, is most strangely changed
from its primary meaning. Who could discern any analogy
between coloured glass and dough, the proper sense of the
word? It comes in fact from *pastus*, food; and the grand
staple of the Southern Italian's dinner has been from time
immemorial the dough of the *gran duro* (the ancient *puls*)
moulded into a variety of curious forms, but all going
under the generic name of *paste*. From the same root
springs the *pâte* of the French, and our own *pasty*, which
varies less from the original. The Italians, when they
revived the fabrication of imitative gems, applied to the
softened plastic material the familiar household word.

Pastes are imitations in glass of the precious stones, and
also of engraved gems, camei and intagli, transparent
and opaque; and the process of the manufacture is the
following : — A small iron case of the diameter required is
filled with a mixture of fine tripoli and pipeclay moistened,
upon which is made an impression from the gem to be
copied. This matrix is then thoroughly dried, and a bit
of glass of the proper colour laid flat upon it. If a stone
of different strata has to be imitated, so many layers of
different-coloured glass are piled upon each other. The
whole is next placed within a furnace, and watched until
the glass is just beginning to melt, when the softened
mass is immediately pressed down upon the mould by

Y 2

means of an iron spatula, coated with French chalk in order to prevent adhesion. It is then removed from the furnace, and *annealed* or suffered to cool gradually at its mouth, when the glass, after being cleaned from the tripoli, will be found to have taken a wonderfully sharp impression of the stamp, but in reverse, whether the prototype be in relievo or incavo. When a cameo is the model, all the undercutting must be stopped up with wax before taking the cast, otherwise it tears away the soft matrix when withdrawn; and on this account camei in paste are never so satisfactory as intagli. If it be wished to imitate a gem full of internal flaws, like the Carbuncle or the Emerald, the effect is produced by omitting the annealing, and throwing the paste, still hot, into cold water.

The process followed by the ancients was doubtless in principle the same, except that it is evident their moulds were taken in a much coarser material (probably in terra-cotta, on which point more will be said in its due place), for antique pastes have a much rougher surface than the modern, and are full of air-bubbles. One singular property, however, distinguishes the ancient manufacture: they are much harder than window-glass, and will scratch it as readily as does a splinter of flint; whereas all modern glass, if coloured, is softer than the white kind. This was due to their different composition, for at present the German glass, made entirely with soda, is much harder (even resisting the file) than the English, into the composition of which enters a large proportion of lead.

De Boot tells us that in his time (1600) rock-crystal pounded was used in the celebrated glass-houses of Venice in making their best articles; and also generally by the Italians in the manufacture of false gems (an art they were then famous for), for which he, sapient old alchemist as he was, gives many curious and valuable recipes. In the present day the trade is transferred to Paris; and it will be seen by a reference to the article *Strass*, how important an ingredient in paste-making crystal is still regarded. Crystal in fact being the purest form of silex, is very superior to the silicious sand now employed, for the sake of economy, in the manufacture. The use of this component can be traced back to the earliest times, for according to Pliny (xxxvi. 66), "some assert that in India glass is made out of pounded crystal, and in consequence no other sort is comparable to the Indian" (CRYSTALLUS). The same fact seems to be implied in the vague expressions of Theophrastus (49): "For if glass be made, as is reported, out of the *hyalites*, this substance also is the product of condensation." And Plato has (Tim. 61 c.) "some substances containing a smaller proportion of water than of earth in their composition as the entire species of glass, and whatever kinds of stones are termed fusible."

Besides this superior hardness, another supposed criterion of an antique paste is the beautiful iridescence, often coating its surface, produced by the oxidising of the glass from the action of the salts of the earth. Another is the porous, bubbly texture, not only of the surface, but of the entire body. The latter peculiarity does indeed distinguish them from those of the modern fabrique, which, when they counterfeit the transparent gems, are pure and homogeneous throughout, being in fact made of what is technically called "pot-metal," glass stained of one colour. These, from the greater fusibility of the composition, usually exhibit a polished surface within the intaglio, not far below that of the real stone whence it was derived. But it must be remembered that an equal hardness would be found in modern pastes, only supposing them made out of the fragments of ancient glass (abundantly procurable about Rome wherever the soil is disturbed), whilst the roughness of the surface entirely depends upon the coarseness of the matrix, which may be obtained at pleasure. As for the iridéscence on which some amateurs set such store, I more than suspect it is often produced by chemical means (fluroic acid). But indeed bits of bottle-glass, after a few years' exposure to the acids of a garden soil, often turn up with surfaces fully as iridescent as the most indubitable antique.

As counterfeits of precious stones, Pastes can be traced back to the most remote ages of antiquity. Herodotus describes the pendants put in the ears of the sacred crocodiles, under the name of λιθινὰ χυτὰ,* "fused gems," a curious yet accurate periphrasis. Such pendants, in blue or green transparent pastes, figure much more frequently than real stones in those elegant and elaborate ear-rings that carry us back to the early ages of Greek and Etruscan prosperity. They also decorate the necklaces of the same period of art. Though their surface be now dulled by Time's corroding tooth, yet doubtless, when first sent out

---

* Epinicus ridiculed the bombastic historian Mnesiptolemus, in a play called after him, for his pompous description of King Seleucus taking his morning draught—ἐν σκύφῳ χυτῆς λίθου (Athen. x. 40). It is evident from the remarks of Theophrastus, cited further on, that the Greeks of Alexander's times had not the art of glass-making. From its rarity, therefore, as well as novelty, a glass bowl was not unworthy of the lord of all Asia. His pupil Hippolochus describes as the crowning piece of magnificence at the wedding of Caranus, after several removes of silver dishes, the entrance of a *glass* plateau about three feet in diameter, inclosed in a silver frame, and piled up with all kinds of fried fish. As with all the preceding courses, one of these then so precious chargers was presented to each of the twenty guests. By Pliny's age, four centuries later, the manufacture was flourishing all over Spain and Gaul (xxxvi. 66). The proficiency of the Celts in the newly-introduced art is convincingly displayed in the admirable composition of the so-called Druidical beads of opaque and transparent colours tastefully combined, and the making of which by exactly the same process was kept up at Murano (Venice) almost to our times.

from the jeweller's at Sardis or Clusium, they displayed more than the polish and the beauty of the Emeralds, Sapphires, and Rubies, whose unattainable rarity they were intended to replace. This notice of early Pastes Fortune has put it in my power to illustrate most strikingly by the description of two extraordinary specimens now adorning the choice collection of Mr. Bale, and acquired from that of L. Fould. The first, an ellipse about an inch in length, is translucent, and streaked with wavy clouds of green, like the Venetian *schmelze*, through which layers of gold aventurine are disseminated in undulations, presenting the most charming contrasts ever achieved in this art. A perforation through its axis shows it was designed to form the centre ornament of a necklace. The other is on many accounts yet more remarkable. The Paste, of the same form as the preceding, is opaque, and composed of wavy patterns in very brilliant colours, arranged as in the *mille-fiore* glass of Venice. But the most interesting circumstance is its retaining the original setting, being encased in a gold collet overlaid with a cable plait in the usual Etruscan taste, and furnished with a massy three-sided shank, affixed to the side with pins, so that the jewel is mounted like a scarabeoid. The weight and elaborateness of the setting manifests the high value set upon the Paste by its original Etruscan owner. Could he have taken it for some rare and unique production of nature ; or was it merely prized as an example of the skill of the Egyptian glass-worker, brought for a precious gift to some wealthy *Lucumo?* * The examination of this skilful combination of numerous patterns, much resembling the old Italian mosaic *a tapete*, seems to throw light upon the obscure allusion to some

---

* The well-known Egyptian idols in blue porcelain have been discovered in the tombs of Valci, mounted in Etruscan necklaces, &c.

such manufacture dropped by Theophrastus, when mentioning "stones brought from Asia," and already quoted under *Sardonyx*, as composed "of earthy minerals, some pulverised, others softened by fire, and so put together." (61).*

The Egyptian glass-workers of the very earliest times produced mosaics minutely finished, and sufficiently small to be set in rings and in pendants to necklaces. The method was ingenious, although extremely simple in practice. A number of fine rods of coloured glass were arranged together in a bundle so that their *ends* composed the pattern wished,—a bird or a flower, exactly as now the makers of Tunbridge ware do with their slips of differently-coloured woods. This bundle was then enclosed in a coating of pot-metal, usually opaque blue glass; and the whole mass being fused sufficiently to run all the rods together into a compact body, was lastly drawn out to the diameter required. In this way all the rods were equally attenuated without altering their relative position; and the external coating, when the mass was cut across, became the ground of a miniature mosaic, apparently the production of inconceivable dexterity and niceness of touch. Each section of the whole necessarily presented the same pattern, without the slightest variation in its shades and outlines. The most exquisite specimen in existence, once the Duchess of Devonshire's, is now in the British Museum. It is a square tablet about an inch wide, exhibiting, upon a ground of the richest blue, a kneeling winged goddess, Sate. In the figure the junction of the threads defies the nicest scrutiny, and produces the effect of an exquisitely painted miniature in the brightest colours,

---

* He also speaks (48) of " earths softened by heat, out of which gems are made, their various colours and combinations being produced by the action of fire."

brought out by the high polish given in the concluding operation to the surface. The back, left rough, clearly exhibits the rationale of the process.

In ' Precious Stones,' p. 221, something has been said respecting the perfection to which the art of making false gems was carried by the Romans. Pliny, in his description of the precious stones, frequently alludes to the difficulty of detecting their counterfeits in glass. As used in the making of drinking-vessels and for other ornamental purposes, he enumerates the following varieties :—" Glass resembling Obsidian * is made for dishes (escaria vasa); also a sort entirely red and opaque, called *Hæmatinon ;* an opaque white also, and imitations of the Agate, the Sapphire, the Lapis-lazuli, and all other colours."† Specimens of all these kinds are met with in abundance amongst other Roman remains even in this country, more particularly the imitation of the Sapphire, a transparent body of the richest blue. The *Hæmatinon* " blood-red " must be the paste, or rather enamel, of that colour, hardly to be distinguished from red Jasper, universally employed in the later mosaics for that tint. Such fragments of the different varieties are collected by the lapidaries at Rome : when cut and polished they are set in bracelets or brooches ; and

* On the strength of this, certain pedants of the last century took to designating all antique pastes, irrespective of their colour, by the name of *Obsidian*—a piece of false learning which greatly puzzles one at first in using the catalogues of the time ; *e. g.* in the Marlborough.

† The most admired, however, was the pure colourless, as much like crystal as possible. Merely because this was the most difficult to produce, any iron in the sand giving the glass a blue tint such as the commoner vessels of the Romans strongly exhibit (MAGNES). Old women used to go about Rome buying up broken glasses in exchange for brimstone-matches, as Martial tells us "sulfurato ramento Vatiniorum proxeneta fractorum," which shows the materials for fine glass (the Syrian sand) were somewhat expensive in Italy.

variegated as they are with bright colours capriciously
arranged, they show like Agates of a novel and superior
species to any from the laboratory of Nature. Translucent
blue traversed with opaque white stripes or patches, and
green similarly diversified with red, were patterns in great
demand amongst the ancients, to judge from the plentifulness of such pieces. But their imitation of the Lapis-lazuli
was the most successful of all; and their reproductions of intagli in this material are the most difficult to distinguish from
the real stone, such is their homogeneousness of substance
and purity of colour. In such is to be seen the finest paste
cameo known (*Townley Pastes*, British Museum), a three-
quarter figure of Bonus Eventus (so inscribed) holding a
cornucopia, already noticed (SAPPHIRUS). The slab is about
8 inches square, and the figure, which is in half relief, has
been carefully gone over with the tool after the manner
of a cameo in stone.* Caylus (ii. pl. 81) has engraved
an admirable Medusa, 4 × 3½ inches in size, in a paste
similarly finished off with the diamond-point; and also a
head of Victory set in a massy bronze ring, which he
describes as a perfect imitation of an Onyx of three layers.
Such counterfeits, so completed, plentifully occur, and often
are most deceptive, so exactly does their surface imitate
the effect of age upon that of the Agate or the Sardonyx.
Yet more puzzling must they have been when recent, for
Martial alludes to the difficulty of finding a real Sardonyx

---

* This is much the largest paste known to me; but it is far surpassed in this respect by another in the Vatican, 'The Triumph of
Bacchus,' which, according to Buonarruoti's fac-simile, is 16 inches
long by 10 deep. Unless, indeed, he is describing the Carpegna
cameo under a mistaken notion as to its nature, for his drawing tallies
with that Sardonyx in dimensions and subject. Such vast pastes must
have been made for panels to coffrets in metal (the *acerra* for instance)
like the crystal plaques of the Cinque-cento.

at the jewellers' of his day, and the fruitless search serving
to kill the morning for his dilettante :—

"Sardonychas veros mensa quæsivit in omni."

Amongst the sorts made in his time Pliny (xxxvi. 67)
mentions the *Murrhinum,* or imitation of the costly murrhine
stone which these very glass counterfeits, were there no
other grounds for the judgment, would teach us formed the
bowls in Agate with unornamented surfaces, such as India
still exports. Glass *scyphi,* exactly reproducing the clouds
and shades of brown and white Agate, largely exist both
whole and in fragments. The still more expensive *vasa
onychina,* cut in Onyx concentrically stratified, and having
their exterior enriched with relievi, upon the dark ground,
in the white superstratum, were emulated in this cheap
material and with admirable taste ; in fact, these substitutes infinitely excel in point of art their more precious
prototypes. A fine example is the elegantly shaped
amphora, 12½ inches high, of the Museo Borbonico, entirely
covered with a network of vines inclosing Bacchic genii,
masks, and goats, in a delicate white upon a dark-blue
ground; also the yet more graceful " Auldjo Vase," an
ewer of the same height festooned with ivy and vine-
branches supporting doves ; both the material and the
execution precisely agreeing with those of the far-famed
Portland Vase.* Such relievi were produced by imitating

* This vase, supposed to have contained the ashes of Alexander
Severus, was discovered about the middle of the sixteenth century in a
sarcophagus deposited under the Monte di Grano, Rome. The subjects
of the decorations have never been explained in a satisfactory manner ;
but one of the panels on the Arca Cypseli, described so fully by Pausanias, has struck me as elucidating the group where a female, seated on
the ground, repels with one hand the advances of a youth, whilst a
huge crested serpent rears itself against him from the other. In the
ancient Corinthian carving, " Peleus is approaching Thetis, 'from whose
hand ' a serpent flies against him." (See p. 364.)

the proceedings of Nature in the creation of the Onyx : the
body of the vase was first moulded in the translucent dark-
blue pot-metal, over which was fused a layer of opaque
white, which was cut away* with the drill and the diamond-point to produce the cameo.   Such a vase (καὔκίον,
caucus in later Latin) is the subject of the epigram (Anthol. ix. 749) :—

> " Why Cupid on the cup?   Enough for wine
> To inflame the heart : why fire with fire combine ? "

Though perfect vases in this style are the rarest of the
rare, yet fragments of such are not uncommon : all that
have come under my observation display the same marvellous taste in the design and equal delicacy of finish.†

One of the most remarkable specimens of this manufacture is the scyphus four inches in diameter found at
Colchester, and belonging to Mr. Pollexfen.   It is embossed
with a chariot-race, exhibiting the Spina of the Circus
with its usual decorations; the names of the respective
charioteers are affixed to their figures.

Glass vessels thus embellished were the "toreumata
vitri" of Martial (xii. 74, xiv. 94), in whose age the word
had lost, with the loss of the art, its proper meaning of a
chasing in *metal*, and come to imply a relievo on a gem or
paste ; in short, our " cameo."   Compared with the fashionable *Murrhina*, these glasses, deriving their sole value from
the engraver's hand, were of but trifling cost and fell
within the means of our necessitous poet :—

* A process to which refers Pliny's expression, " aliud argenti modo
cælatur."

† This was the "vitrum fabre sigillatum," deemed not unworthy to
figure on the table of the wealthy Byrrhæna amongst "crystallum impunctum argentum alibi clarum et aurum fulgurans et succinum mire
cavatum in capidas ut bibas . . . . gemmas formatas in pocula vini
vetusti" (Apul. Met. II.).

" Cheap vulgar chasings we in glass alone
That dares to counterfeit the precious stone ;
Yet far superior to the crystal rare,
Not boiling water flaws our substance clear."

.   .   .   .   .

" See the sweet groups which on the goblet smile—
Fruits of the taste that drinks the tepid Nile.
Ah ! whilst he strives to add another grace,
How oft the artist spoils the finished vase."

The glass-workers of the Lower Empire, stimulated by
the patronage of the luxurious (like the Emperor Tacitus
who, records Vopiscus, was an enthusiastic admirer of their
productions),* invented novelties in the art as far surpassing
the last-noticed pieces in technical skill as they fall short
of them in artistic merit. Of such, the most elaborate
example is the one now in the possession of Baron Lionel
Rothschild. It is ascribed to the third or fourth century :
the substance is of a pale ruby colour by transmitted light, •
and of a pale opaque green by reflected light. It is semi-
oviform in shape, and decorated with figures, vines, &c., in
full relief, in some cases completely detached from the
glass itself. The mass of the bodies of the four principal
figures is hollowed out from the inside ; the other decorations are fixed on, and are all of them hollow so as not to
interfere with the colour of the glass. The whole has been
carefully cut and polished with a tool, like cameo-work.
One of the figures represents an aged man whose limbs are
entangled with vine-branches ; perhaps Lycurgus king of
the Edones in Thrace, who was driven mad for persecuting
Bacchus and his followers. To his right is a woman
prostrate, behind her a Faun dancing, and holding in his
left hand a *pedum,* in his right a stone which he seems
about to throw. To the left of Lycurgus is Pan, followed

* " Vitreorum diversitate atque operositate vehementer est delec-
tatus."

by a dog or panther: behind him is Bacchus pointing to Lycurgus with his right hand, and holding a thyrsus in his left.

The workmanship of this remarkable vase belongs to the class called by the ancients *diatreta*, which seem to have been made by a special class of workmen called *Diatretarii*. This style of drinking-cup belongs, to judge from Martial's exclamation, to the highest extreme of luxury (xii. 70).

> " O quantum *diatreta* valent et quinque comati."

The strangely opalescent paste composing its body refers its production to the Alexandrian fabrique, to be mentioned further on.

These imitative Onyx-vases obtained a special name from the boldness of their pretensions, being called 'Calices Audaces,' or 'Impudent ware.' This is established by the title 'Calices Audaces' affixed by Martial himself to the epigram quoted last page, in which also the very epithet is repeated:—

> " Nos sumus *audacis* plebeia toreumata vitri."

And again (xii. 74) he compares the effrontery of the manufacture with that of himself in presuming to offer so trifling a present to a friend as a couple of these vases:—

> " Dum tibi Niliacus portat crystalla cataplus
> Accipe de Circo pocula Flaminio :
> Hi magis *audaces* an sunt qui talia mittant
> Munera ? . . . .

Where 'audaces' following 'pocula' proves the former to be a generic title.

Finger-rings formed entirely in glass are known; the shank of a twisted pattern imitates in colour the Agate; they bear a mask in relief, usually in green, representing a Plasma cameo. One fell into my hands when at Rome, recently found, which exactly corresponds with that given by Caylus (ii. pl. 89). From the fragility of the substance, perfect examples are extremely uncommon.

A singular variety of the antique paste, something between a mosaic and a cameo, is afforded by those disks, the size of ring-stones, the ground imitating Lapis-lazuli, and inlaid with other pastes forming a pattern, in very low relief. I have seen only two (Hertz), a vine-leaf and a parrot, which, being considered unique, sold for 10*l.* each: and the former of these is engraved in one of Caylus' plates, as a marvellous example of the art.

Under this head come most appropriately the glass disks found stuck in the plaster (before it set) which closes up the tombs in the Roman catacombs. They are for the most part the bottoms of cups, of which the upper part has completely mouldered away by age, placed with the corpse to contain the sacramental elements. Several, however, were indubitably no more than medallions manufactured to be worn for pendants by the poorer classes (who could not afford their prototypes in gold so fashionable under the Lower Empire), for they are encircled by a moulded border and having a loop attached for suspension. Within their substance they contain rude designs cut out of a stout piece of gold-leaf laid between two plates of glass, afterwards fused together and so forming one solid body. Usually the subjects are busts or full-lengths of the Saviour, or the Apostles, with appropriate legends; a few have the figure of the reigning emperor. As medallions they seem to have accompanied the interment partly as amulets to guard the repose of the dead, partly to mark the date, like the coins deposited in the sepulchres of the ages preceding.

That the Christians, however, actually used drinking-glasses adorned with religious emblems, and this at an

early period of the Church, is manifest from Tertullian's
apostrophe (De Penitentia): "Where is the lost sheep?
let the very paintings upon your drinking-cups (calicum)
come forward." But the invention of this species of decoration goes much further back, as is manifest from the
exquisite portrait of a child, worthy of the best times of
Roman art, now in the Townley Collection. Another piece
figured by Millin (Gal. Myth. pl. 33) has full-lengths of
three girls, Gelasia, Licoris, Comasia (whose names plainly
indicate their profession), portrayed as the Graces, with
the motto in Græco-Latin, "Piete Zesete multis annis,"—
Drink: may you live many years.

"Mosaic pavements," observes Pliny (xxxvi. 64), "have
been driven from our floors and have migrated to our
ceilings: and are made of *glass*, a new invention this; for
Agrippa in the baths he built in Rome used terra-cotta
decorations painted in *encaustic* in the heated chambers, for
the other parts employing stucco-work, whereas he would
certainly have made his ceilings of glass had the invention
existed in his time." The first species of mosaic, the *lithos-
troton*, was made, as its Greek name denotes, out of minute
bits of marble "*parvolis crustis,*" and the earliest example
known in Italy was the floor of the Temple of Fortune at
Præneste built by Sulla. Cav. Barberi, director of the Papal
Mosaic-studio, once informed me that all the finest antique
mosaics, such as Hadrian's Doves, are entirely composed of
cubes of natural marbles of divers colours: afterwards bits
of glass were introduced for the brighter tints (of which a
striking example may be found in the Circencester pavements) but mosaics of the Lower Empire of which the
earliest remaining specimen is the ceiling of Sta. Costanza,
are entirely composed of cubes of coloured glass roughly
broken from the mass.

Hadrian sent his brother-in-law Servianus as a present

from Alexandria, records Vopiscus (Sub Saturnino), two cups in *schmelze* glass, "calices allassontes, versicolores," offered by the priest of the Serapeum (answering to the Patriarch of Alexandria of later times) as specimens of the manufacture for which his city had long been celebrated;* and in truth very curious is this paste, for, though a bright green, it turns by transmitted light to a ruby red.† Recommended by such a property, regarded as miraculous in the ages of barbarism that followed, a lump of schmelze lay from time immemorial upon the famous " Escrin de Charlemagne," that grand ornament of the Trésor de S. Denys, and was exhibited to worshippers as the very gem of King David,—as was certified to the faithful by the legend cut upon its setting : " Hic lapis fuit Davidis regis et prophetæ,"—and a fitting companion to the illustrious Beryl with the head of the Queen of Sheba (now restored to Julia Titi) which capped that wondrous fabric of the Carlovingian goldsmith. Pliny also mentions draughtmen made of pastes of many varying tints, "pluribus modis versicolores."

These *latrunculi* are so curious both in material and in ornamentation, that they demand a separate notice before we quit the subject of antique pastes. They are disks of glass of different colours, about an inch in diameter, bearing in relief the head of a deity, or some other emblem. They exist in large numbers, one proof amongst the rest that they are the latrunculi, or men used in the very popular game of backgammon.‡ That these men were usually made of

* To declare the incredible wealth of Firmus, the Alexandrian merchant, who often boasted he could "keep an army on his paper alone," Vopiscus mentions he had his palace cased all over with plates of glass (vitreis quadraturis) cemented with bitumen.

† I have seen amongst Mr. Webb's antique glass a large paste, an intaglio head, in this remarkable variety.

‡ A name signifying in Welsh "a little battle"—a singular coincidence with the primary meaning of *latrunculus*, "a little soldier."

(c)

glass appears from many passages of the classics. Ovid has the "vitreus miles," and Martial "vitreus latro." For these, as something grander, Trimalchio substitutes gold and silver coins in the two colours marking the sides:—

"Calculus ut gemino discolor hoste perit"—

his dice being made of rock-crystal. Later, luxury progressing made the men in precious stone, probably the Onyx, for Martial (xiv. 20) has—

"Insidiosorum si ludis bella latronum
Gemmeus iste tibi miles et hostis erit."

"If thou the tricky game of tables play,
Alike in gems soldier and foe survey." *

Whoever desires to understand the real nature of this game, often erroneously supposed to be chess, let him read (and make out if he can) the minute account of a grand match played between the Emperor Zeno and the poet Agathias, and immortalised in an epigram of the latter's (Anthol. ix. 482).

The art of making Pastes was never lost; it afforded too great facilities to fraud of the most lucrative kind ("neque est ulla fraus vitæ lucrosior," says Pliny), ever to have fallen into disuse in the jeweller's *atelier*.† Pliny alludes

* The famous chess-board presented to St. Louis by the Old Man of the Mountain is of crystal, the pieces themselves in agate : the whole on a gigantic scale worthy of its history. That very opposite to the chivalrous saint, the Eighteenth of the name, commonly used this board, until two of the pieces were stolen, when in disgust he deposited it in the Museum (Hôtel de Cluny).

† A remarkable instance, the great Ruby of Charles the Bold, set in a golden rose, found amongst the other spoils of Granson, has been already quoted ('Precious Stones,' p. 232). Again in that very interesting historical relic, the Darnley or Lennox jewel, supposed to have been made in memory of Lennox, Regent of Scotland (murdered in 1572) for his widow Margaret Douglas, the large Sapphire in the centre is unmis-

to the numerous treatises on the subject then current. Heraclius, who flourished, as his editor Raspe thinks, in the 7th century, and appears to have been a Spanish Goth, to judge from certain provincialisms of his barbarous dialect, wrote a manual 'De Artibus Romanorum.' Amongst others, he gives a recipe for making Pastes, but has unfortunately chosen to clothe his meaning in such obscure and corrupt hexameters that a translation is altogether beyond my powers, and therefore I can only present my readers with a transcript of the original to exercise their sagacity upon :—

"*DE GEMMIS QUAS DE ROMANO VITRO FACERE QUÆRIS.*

" Sic a Romano poteris conficere vitro
Splendentes pulcros generis cujusque lapillos,
Ad modulum lapidis *cretam* tibi quippe cavabis
Hic ponas vitrum per quædam frusta minutum,
Hanc ergo facile poteris hac arte parare :
Subtiliter quædam circumvolvatur arundo
Duc dum durescit cum virga fortuis hæret,
Tunc ipsi virgæ superimponatur utrinque
Et circumposito teneatur virgula vitro,
Atque cavo tectam penitus tunc insere cretam.
Ignifit vitrum cum sit penitus liquefactum
In fossam lato fulgenti comprime ferro ;
Quo vesica tibi quo læsio nulla supersit." *

---

takably a paste, whether original, or substituted by some later and necessitous heritor. It once belonged to Horace Walpole, and was purchased at the sale of his collection in 1842 for Her Majesty.

* A portion of this is intelligible enough : the matrix was to be made in pipe-clay, *creta*, and covered with the glass broken small ; but for what purpose was the reed introduced ? When melted the paste was to be pressed with a flat, polished, iron to prevent bubbles and flaws. Pliny also has obscurely alluded to this employment of *creta* in paste-making : "annulare (white paint) fit ex creta ad *istas* vitreas gemmas ex vulgi annualis :" for so the sense requires it to be corrected from the present reading, "*admixtis* vitreis gemmis" for a *white* paint composed of *coloured* glass mixed with clay is a contradiction in terms.

In the crowns of the Gothic nobles discovered at Guerrazar false Emeralds and Opals may be detected amongst the real stones, evidently intended to pass for such, and particularly to replace the Emerald, which appears to have been then excessively rare in Spain, though the court-jeweller had abundance of the most beautiful Sapphires at his command. ('Precious Stones,' p. 311.) .

But what is most strange, that masterpiece of Celtic *bijouterie*, the far-famed Tara Brooch, exhibits specimens of this art, in the shape of four female heads in red paste, all from the same mould, set in the handle of the jewel, and clearly belonging to the same style as the rest of the ornamentation.

Antique pastes are often to be met with retaining their ancient bronze settings. The most interesting example of the class I have met with is in the Waterton Dactyliothea, and was found on the finger of a skeleton in an Etruscan tomb. The ring had been strongly gilt, and was set with a good paste, a long pointed oval, moulded upon an intaglio, in a very archaic style, of Bacchus and Ampelus. But pastes seldom appear in rings of gold; just as might have been expected, for they were by their nature the substitutes for real stones to the lowest vulgar and to slaves. Pliny mentions the glass gems from the rings of the populace, in his above-quoted notice of the white paint thence called " annulare." A paste cameo, however, a seated sphinx, a good imitation of the Sardonyx, set in a massy gold ring, once came in my way : but the cameo had beyond all doubt been palmed off upon the ancient buyer as a true gem and the same may be supposed the case with the counterfeited Carbuncle intaglio, a vase, elegantly set in gold, also amongst Mr. Waterton's. Similar frauds remind us of the jocular punishment inflicted by Gallienus upon the jeweller who had thus taken in his Empress, Salonina, by selling her

" vitreas gemmas pro veris." She having demanded that.
a condign example should be made of the at once cheat and
traitor, the Emperor sentenced him to be exposed to the
lions in the arena. The poor wretch stripped naked was
thrown in, the door of the den was raised, when out strutted
a capon; and the culprit got off with the fright alone.
Gallienus proclaiming by the voice of a crier that the
cheater had been cheated: " imposturam fecit et passus
est."

The finest example of an antique piece of imitative
jewelry that I have ever seen was found amongst other
Roman remains at Shefford, Bedfordshire, and is now in
the museum of the Cambridge Antiquarian Society. It is
in bronze thickly gilt, about eight inches over, and formed
by three concentric circles connected by pieces radiating
from the centre, and set at all the intersections with large
paste Emeralds and Amethysts of beautiful colour, retaining
their first polish. Its pattern is unique for a Roman fibula,
but is identical with that of the head of a monster ring,
once shown me, a present from the Great Mogul to an
Austrian envoy some two centuries ago: the concentric
circles whereof contained every gem known to the Hindoos
except the Diamond.

The best antique intaglio Pastes that ever came before
me in their original mountings were; one found near Rome
in the spring of 1850, its subject the Town of Troy upon a
skilful imitation of the banded Agate, retaining its antique
massy bronze ring almost entire; and another, the bust of
Abundantia in imitative lapis-lazuli of large size mounted
in gold for a pendant (Eastwood). A singular recognition
of the prototype of a paste, the genuineness of which is
beyond all question, once delighted myself. Amongst the
Praun Gems was a paste, of Thalia seated, holding a mask
on her knee. At the sale of the Uzielli Collection in 1861,

I recognised its parent in a fine Sard, then deficient in the
lower part, although perfect when the paste was cast from
it some eighteen centuries before.   In consequence of my
information the gem was secured by the then owner of the
original: a coincidence perhaps unknown in the history of
gem-collecting.*

As soon as the Glyptic art revived in Italy, the former
makers of false stones now set to work to reproduce intagli
in their own material, and such occasionally are to be met
with in the jewelry of that age, by which their date may be
positively ascertained.   For the most part these first essays
are coarse, far inferior to the antique, and done in blue
glass to imitate the Sapphire.   At the same time, if a good
antique paste fell into an engraver's hands, existing specimens show that, after repolishing, he passèd it off for a
real stone upon the enthusiastic but uncritical lovers of
antiquity amongst his patrons.   I have seen a false Garnet
representing a Bacchic Feast with several small figures in
the best style, mounted, in virtue of its supposed value, in
a magnificently enamelled gold ring of the times of Clement
VII.   The paste of recent manufacture most in use during
that period was an opaque kind, a dull grey marbled with
green and red, badly counterfeiting the Agate, but hard,
taking a good impression of the matrix, and having much
the look of a real stone, especially when polished.   This
false Agate is particularly mentioned by De Boot.   But all
that was done in this line was of small account until the
art was taken up by the Regent Orleans, under whose
patronage it rapidly attained to perfection and celebrity,

---

* Similarly the Berlin Cabinet possesses amongst Stosch's gems a
paste taken from the Taras on the dolphin, a Beryl existing in the
Praun.   Winckelmann, who has described the paste in terms of the
highest admiration for the beauty of the design, was not acquainted
with this fact.

its productions far excelling the best left us by the ancients in this particular walk of art.

Clarac gives the following history of the Orleans Pastes : " Having engaged (1691-1715) the services of the cele- brated chemist Homberg, and assisting him with his own hands in his operations, in a laboratory established within the Palais Royal, the Regent made him reproduce in glass- pastes all the gems that he himself had collected; and besides these, a large number selected from the royal cabinet. It is said that he made six complete sets of these Pastes, one of which Clarac himself possessed, the bequests of M. Gosselin of the Academy. In the keeping of the latter it had been for many years, and always was regarded as one of the six original sets proceeding from the Regent's own laboratory. It had, however, been augmented by several additions of Pastes, probably made by Clachant and Mdlle. Falloux, who had been instructed by Homberg in the manufacture, and set up as dealers in its productions.

" These Pastes of the Regent's are in a very fine glass and also in enamel, and exactly reproduce the colours of the original stones. It is evident that they were manipulated with the utmost carefulness; the substance is very dense, and free from flaws and air-bubbles. The intagli in it are clear, lustrous, and internally *polished*, a result extremely difficult to obtain.* When held up against the light the transparent kinds amongst them produce by the richness of their colours precisely the effect of real stones. Some of them indeed, particularly the *Surdoine* (brown Sard), have been imitated better subsequently, so far as the tone of the colour is concerned, but nevertheless, in spite of the modern improvements in the art of glass-making and enamel, and in spite of the advance of chemistry, it is very doubtful if

---

* And which indeed no pastes of any other fabrique ever exhibit.

better pastes than those of the Regent could be produced in our times."

The new process rapidly diffused itself throughout Europe; and when Goethe visited Rome in the last quarter of the century (1786-9), he found the making of such Pastes a favourite occupation with the dilettanti residing there. Even at the present day the Romans display extraordinary skill in the art: I have seen some of their Pastes, especially of the opaque sort, like the Sardonyx, that could only be distinguished from the true gem by the file. To baffle this mode of detection, the ingenious device is resorted to of backing the paste with a slice of a stone of the same colour. When set in a ring the junction is concealed, and the back withstanding the test of the file enables the composite to pass muster for a real gem, adorned with a precious engraving.* The same method has been long, and still is, largely adopted for counterfeiting the coloured precious stones, the Ruby, the Sapphire, and the Emerald. A Paste of the proper colour, backed by a piece of crystal facetted so as to impart the requisite brilliancy, is often sold to the unwary for a gem of the first class, nor is the fraud suspected until the wear of some months begins to tell upon the vitreous surface of the upper layer. Pliny mentions a similar trick as practised by the lapidaries of his day in the case of the Sardonyx; its three several strata being made up out of three different stones, each of the best colour for its position, and the whole cemented together with Venice turpentine, still preferred for the purpose on account of its perfect transparency.

* As an instance Clarac mentions having been shown a paste taken from a gem of Marchant's, and actually retaining his usual signature, which had thus been metamorphosed into an antique Sard. With satisfactory vouchers of its having just been dug up at Otranto, it was sold at an enormous price to a Neapolitan duke, an enthusiastic but not experienced amateur.

Wedgwood's seals and camei close the series of modern pastes : but in their material they differ from all hitherto described, being not glass but porcelain. This substance is far superior to paste in hardness, even striking fire with the steel. There are two kinds, one imitating black Jasper, the other Lapis-lazuli. Their subjects are moulded upon antique originals, their impressions are remarkably sharp and clear, although wanting the polish of the Italian glass-pastes. It is therefore impossible to mistake them for true gems, although they rival the hardness and durability of the latter. But Wedgwood's excellence lies in his camei, many of which are from Flaxman's designs : others are reduced copies of the antique. The figures stand out in opaque white upon a turquois ground, of a charming shade, which the numerous imitations now-a-days in the market are totally unable to reproduce. In his genuine works it will be discovered that all the reliefs have been elaborately worked over with the wheel, thus following the antique mode of procedure in their best camei in paste. The finest specimen of such, the Portland Vase, he copied with extraordinary success, though somewhat failing in the dark-blue of the ground. Of this he produced fifty facsimiles, at fifty guineas each ; a price, however, that did not cover the expense of the workmanship, so great an amount of skilled labour did it engross. His other camei appear in as many varieties as their ancient prototypes : some in the form of small medallions for setting in brooches or bracelets; others decorating the exterior of vases and tea-services ; and a third kind, the most important, large plaques for hanging up as pictures. Imitations of all the above are now brought out in countless swarms; but they are miserable parodies upon Wedgwood's originals, manufactured merely to sell, after the fashion of our days of "art for the million"; the ground a cold pale blue, the reliefs, full of cracks and flaws, are merely stuck on, often imperfectly, without any attempt at a finishing polish.*

I shall conclude this notice with an observation which will be extremely unpalatable to most amateurs—that, amongst the myriads of pastes passing current for antique, hardly one in a hundred is really genuine. This opinion is based upon the following grounds, suggested to me by experience : in the handfuls of gems perpetually brought in for sale to the Roman *antiquari* by the peasants, just as they find them in turning over the ground of their vineyards and gardens on the ancient site of the city, pastes hardly ever occur without some portion of the original bronze ring adhering to them : the *loose* gems being invariably stones which from their base quality and careless execution had evidently belonged to the poorest classes. Besides, as these valueless imitations were never worn by people who could afford rings of gold or even of silver, there is no room for the saving clause that they might have been drawn from their settings and thrown away, as was evidently the case with real gems when in barbarous times the jewels they adorned were melted down for the sake of the metal. Again, whoever has seen a paste in its original setting must be aware of the impossibility of extracting it without breakage. Had paste intagli been as common in Roman times as they are in dealers' trays, they would have formed the majority of the gems turned up in the soil covering ancient towns, whereas the direct contrary is the case. All pastes, therefore, that appear never to have had an ancient setting, ought to be regarded with the utmost suspicion.

* Tassie, a Scotch sculptor established in London (1760–90) brought out a series of pastes to the number of 15,000. They include the entire Stosch gems, with a selection from all the other cabinets then existing. Raspe was employed to write the catalogue, illustrated with etchings by Allan of Edinburgh, slight though well-drawn : the most serviceable work the student of the Glyptic art can possess.

The abundance of pastes, all passing for antiques, on the accumulation of which so many amateurs pride themselves, is perfectly astounding, and reads an amusing commentary upon Ovid's dictum—

"In quod volumus credula turba sumus."

Many dilettanti exhibit them by hundreds at one and the same time, and must therefore cherish the belief, if they wish to support their authenticity, that the old *vitriarii* spent their lives in producing these imitations by myriads merely for the purpose of sowing them broadcast in the earth to turn up again for the delectation of future ages. The rate at which they are sold depends entirely now upon the credit of the cabinet to which they may belong: sometimes I have known them go for two or three shillings per dozen, and those, too, good examples in their way.

To account rationally for this plentifulness, it must be borne in mind that for two centuries (before the Regent) the Italians were largely manufacturing these articles, their material being the hard glass then commonly made, and their moulds coarse, as their false Agates plainly show; it is therefore hard to perceive how their fabrications can be discerned from the ancient, being produced under conditions so similar.

As for the fine specimens that began to appear after Homberg had so greatly improved the process, their polish and their softness forbid their entering at all into the question. And under the same head may be classed all that followed; even the amateurs, whom Goethe speaks of, having so far made themselves masters of the technical processes when once invented, as to be able to bring out pastes free from bubbles, and exhibiting a clear impression of the matrix. Modern pastes betray themselves by one almost certain mark of their origin, when they retain at the back the thin projecting rim of glass formed by the portion superfluous and overlapping the circumference of the casting-box, a convincing proof they were never set at all, or even intended for setting; and such ought to be put down to the account of the amateur glass workers of the last century as merely made for the cabinet.

An interesting lot of the raw stock in trade of some old *vitriarius* once passed under my examination. It amounted to above two hundred masses of glass, in form and size resembling gooseberries of different kinds: a shape given them, as it was to the blanks in the contemporary coinage, to enable them, when semi-fused, better to take the impression of the mould. Some were of rich colour, others ingeniously divided by an opaque band intended in their finshed state to imitate banded Agates. The whole stock, including a remarkably fine bust of Jupiter in relief, (emerald paste) and a few rudely engraved Sards and Garnets, was said to have been found in one deposit in the vicinity of Naples. Could this account be depended upon, such a collection of the raw material would have greatly illustrated the process used in this curious art: but unfortunately no reliance whatever is to be placed upon the tale of the *provenance* of any antiques brought from that country, however well authenticated it may appear to the inexperienced. But a long time back an instance beyond all suspicion came under my observation, of a paste prepared for the matrix. It was of a dark, translucent blue, lenticular in shape, and found together with a carnelian ringstone, not engraved, and a silver ring set with a red Jasper bearing a rude intaglio, all deposited under a large stone amongst some ruins of the ancient Isca Silurum.

This single relic supplies in itself a reason for believing that the art of making paste intagli was then carried on in Britain, as we know was the case with that of bead-making, and corroborates the opinion expressed by Wright (in his 'Celt, Roman, and Saxon.') that the rolled lumps of beautifully coloured glass found in the shingle on the Brighton beach are the relics of some Roman-British manufactory once occupying the site, and ought not all indiscriminately to be accounted the mere harvest of the lapidaries who annually sow the sands with broken bottles to reap such lucrative returns.*

The ancient process of *glass-making* was simple in the extreme, and the result of a lucky accident. Some Phœnician traders, returning from Egypt with a ship-load of natron (native soda, the soap of the ancients), chanced to land at the mouth of the river Belus, close to the city of Ptolemais in Palestine; and not being able to find any stones on that sandy and muddy beach wherewith to prop up their caldron over the fire, they used for the purpose some lumps of natron out of the ship. Afterwards finding, in the ashes of their fire, streams of a novel and shining substance, they repeated the experiment, and so discovered the components of glass. For many ages this strip of the coast, not more than half a mile in extent, supplied the world with sand for this manufacture; but in Pliny's age it had been ascertained that the sand from the mouth of the Volturnus (Neapolitan) was equally good. Sidon was the seat of the manufacture for a long period. Alexandria succeeded for works of the more artistic kind,† and main-

* This opinion has lately received strong additional confirmation from the inspection of what was shown me for a curious pebble picked up on the sands at Tenby. It certainly much resembled a transparent Calcedony, containing two petrified, striped, snail-shells, but proved to be an unfashioned mass of ancient British glass, rounded into a pebble-form by the action of the sea.

† Here was invented the most elegant and durable representative of value ever devised, the glass money issued by the Fatimite sultans dating from the tenth century. It consists of thick disks of green glass, bearing a legend in letters raised in red enamel.

tained her reputation long down into the Middle Ages, producing such remarkable enamelled pieces as the Cup of Charlemagne,* so long treasured up in the Madeleine, Châteaudun. It sounds oddly to a modern ear to find Pliny alleging, as a proof of the proficiency of the Sidonian workers, that they actually made *mirrors* in glass; but in his time, and long after, *speculum* metal, a particular alloy of bronze, was considered the only fit material for the purpose; and even in the 15th century the Italians preferred mirrors in polished steel, an example of which exists in that said to be Lucrezia Borgia's in the Soulages Collection. One Sidonian artist at least has contrived to perpetuate his name, for many handles of vases are known stamped with ARTAS. SIDON., and the same repeated in Greek characters.

In the old method they employed, besides the sand of the Belus, "white pebbles" (quartz) pounded, with a proportion of copper, and melted the mixture with a fire of dry resinous wood.† But later they had learnt to whiten their glass with the mineral *magnes*, evidently our manganese, "magnesia vitriariorum." The first melting was called *ammonitrum*, from its ingredients—sand and one-third of *natron*: this was melted again and purified before being worked into vases. The substance was elaborated in a variety of ways, "some being fashioned by blowing, some polished on the lathe, others engraved after the

---

* In reality made towards the end of the twelfth century, as the fine inscription covering its surface announces. The same city was the source of the Venetian manufacture: the earliest glasses of that fabrique which go back as far as 1490, are gilt and enamelled in precisely the same style. (See J. C. Robinson's 'Memoir on Venetian Glass' in the Soulages Collection.)

† "*Continuis* fornacibus sicut æra," in a succession of furnaces like copper, without being allowed to cool; or it may be in a reverberatory furnace like that now used in copper-smelting; for in his notice of *cadmea* he mentions the *camara* as well as the *latera* of the furnace, showing that it was arched over.

manner of silver-plate." Glass, remarks Pliny, had superseded gold in the form of drinking-vessels at the tables of
the great; but the gain in point of economy was but little,
since these were manufactured with such elegance that
Nero paid six sestertia (60l.) for a pair of small two-
handled vases ("pterotæ"), doubtless belonging to the
same class of artistic productions as the Portland and
similar rarities already noticed.

If there be any truth in a strange story alluded to by
Pliny, as "more popular than well authenticated," a secret
had been once discovered in the art which there is small
likelihood will ever be regained. "It is said that in the
time of Tiberius was found out a way of tempering glass
so as to make it malleable, but that the inventor's entire
establishment was exterminated, for fear the value of bronze,
silver, and gold should suffer diminution in consequence."
It must be remembered that Pliny was born under this
emperor, and would not have alluded to the report had it
not been a matter of public notoriety. Petronius, Pliny's
senior by some twenty years, makes his Trimalchio, an
amateur in virtue of his wealth, (our 'railway millionaire')
give a humorous version of the same tale:—"I prefer
glasses; others do not. If glasses were not so brittle, I
would rather them than gold; as it is, they are of little
value. Yet there was once an artist who made a glass that
would not break. He was admitted before the emperor
with his present. After the latter had admired it, he
requested Cæsar to give it him back again, and then
dashed it upon the pavement. The emperor could not
help being frightened almost out of his wits; but my man
picks up the bowl from the ground, and lo! it was only
bruised, just as if it had been of copper. Thereupon he
takes out a little hammer, and leisurely makes it all right
again. Having done this, he thought himself already at

the summit of his wishes, especially when the emperor asked him, 'Does any one else know of this mode of tempering glass?' Now mark me. As soon as he had answered 'No,' the emperor ordered him to be beheaded then and there; for if his invention had become general we should look upon gold as so much dirt." There may be something at the bottom of the legend, and it may be explained by supposing the substance not a true glass, but some composition representing the coloured kind in which the Roman scyphi were usually made. A ring found in a mummy-case once came into my hands, exactly resembling amber-coloured glass, but which was as flexible as india-rubber, being probably moulded in some resinous composition. Now a bowl in a similar substance would exactly resemble glass, and yet be as ductile as the softest metal, even as thin pewter. Supposing such a counterfeit to have been passed off upon Tiberius for a new discovery in glass-making, the inventor's recompence for the trick, when detected, would have been sharp and summary, his patron not being one to let off an offender so easily as did the jocular Gallienus. Moreover, a few years ago there was exhumed from a tumulus in Ireland a perfectly pre-served bowl in Amber, of so large diameter that it was considered impossible it could have been hollowed out of a single block. It was ascertained at the time, by the ingenious discoverer of the relic, that bits of Amber boiled in turpentine can be easily reduced to a plastic mass, can be kneaded together, and moulded into a consistent whole; and this was supposed the true method by which this gigantic *scyphus* had been elaborated. Under "Succinum" will be found some ancient recipes for the purifying of Amber, and also for the staining it to imitate certain precious stones. This shows that the Roman lapidaries were fond of experimentalizing upon the capabilities of this singular mineral. In the description of Byrrhæna's feast, above quoted, *capides* hollowed out of Amber make a conspicuous figure. The Romans could scarcely have afforded solid pieces for the formation of flagons, however small, but their value was enhanced by representing them to have been so formed.

The ancients, though such proficients in the making ornamental glass, and in their manipulation of the *metal,** had, so far as can be discovered, no secrets not possessed by the modern trade, except those unattainable now, superior skill and finer natural taste. The same reasons account for their achievements in the Glyptic art, although, if any faith were to be placed in the strange recipes which Heraclius pretends to have preserved, they had found out *compendia* for annihilating every difficulty. That barbarous writer, however, was like the rest of the Mediæval philosophers, who had by their education lost all power of distinguishing truth from falsehood, and who, having determined, from fancied and fanciful analogies, that certain means *ought* to produce a certain result, proceeding according to a woman's logic, incontinently set down such means as actually effecting the object desired. Nevertheless his recipes are in themselves so curious that I shall transcribe a few of the most remarkable, although strongly disposed to consider all as little better than the dreams of some monkish craftsman too lazy to try the experiment of their truth. In the frequent introduction of earth-worms into his *menstrua*, there may lurk some reminiscence of Solomon's worm *Samir*, whose blood dissolved all gems—that fable of the doting Rabbis, engendered upon the employment of the *smyr* (emery) by the practical Greeks.

---

* Caylus notices that the Parisian trade were particularly struck in the case of certain Roman urns at their being made without the *ponty* or mark of the adhesion of the blowing-tube.

(G)   2 A

HERACLIUS ON THE MEANS OF SOFTENING GEMS.

The notion entertained by many, and which indeed involuntarily forces itself upon the mind on examining the facile execution of their glyptic works, viz., that some secret for softening the gems beforehand was known to the ancients, derives a certain degree of confirmation from the following recipes of Heraclius, who wrote at a period (the 7th century) when the traditionary processes were still kept up, though in a faint and feeble existence. If we could only believe his positive assertion in his prelude, that he had himself tested all his recipes, the whole mystery would be solved at once; but his promises are so extravagant as to make one more than suspect that, instead of disclosing ancient trade-secrets, he is merely passing off in their stead chimerical theories of his own never reduced to practice, after the common manner of Mediæval quacks.

Extracts from a MS. of the 13th century, formerly in the Library of Trin. Coll., Cambridge,* published in Raspe's Treatise on Oil-Painting.

"*Of engraving upon Glass.*—Ye artists who wish to engrave glass handsomely, now will I disclose to you a method exactly as I myself have proved it. I collected fat earth-worms turned up by the plough out of the ground; and at the same time *the art* useful for such matters directed me to get vinegar, and the hot blood out of a big he-goat: which I skilfully fed upon strengthening herbs for a short time, when kept tied up in-doors. With the hot blood I then mixed the worms and the vinegar, and so anointed

---

* Stolen thence since 1840, and subsequently bought, under highly suspicious circumstances, by the British Museum authorities, who have steadily shirked inquiry, and refused restitution.

the whole of the bright shining bowl; which being done, I essayed to engrave upon the glass with the hard stone known by the name of Pyrites.

" *Of the engraving in Precious Stones.*—Whoso desires to attack with the steel the noble gems which the princes of the Roman city, they who formerly occupied her lofty palaces, loved far above gold, let him receive, for it is extremely precious, a device which I have discovered by deep thought. I got myself *uricia* (urina ?), and also blood from a big he-goat, fed upon grass for a moderate time: which being done, with the *hot blood** I engraved the gems, according as the author Pliny directs—and

" *How Stones are engraved.*—Take a he-goat that has never paired, and put him in a cask for three days. Afterwards you must kill or bleed the goat, and mix his blood with his urine, and so steep the stone in it for one night; and afterwards either press it into a shape, or carve it as you please. In order to make it beautiful (*i. e.* to polish it) make yourself a plate of lead, and sprinkle this with white flint ground small like pepper, and rub the stone upon it until you have smoothed down its roughness. Then tie up some of the same pounded flint in a woollen cloth, and rub therewith the corners you were unable to get at with the leaden plate. Then that the stone may recover its original lustre, get walnut-oil, and rub it therewith. Furthermore you must rub it with a waxed cloth, so that it may shine, and cease sweating.

" *How Crystal may be engraved.*—Take crystal and wrap it up in a linen cloth steeped in the sweat of a she-goat, and then cover it up with the cloth itself in cow-dung, and so cut it with a knife as you may wish, but yet with

---

* "Calefacto sanguine," judging from the context, should be " calefactus sanguine," the *gem* being heated or boiled in the blood.

2 A 2

caution.   After you have done, throw it into cold water.
Lastly rub it on a plate of lead with flour or bran.

"*How Glass is engraved.*—The teats of a she-goat are
flogged with a stinging-nettle, and pressed with the palms
of the hands so that the milk may come down in them.
Afterwards the milk is pressed out into a vessel, and in it
let the glass be placed for one night, together with the
steel tool by which it must be engraved.   But when it is
required to be used, let it be warmed again to the same
degree of heat it had when it was first milked, and let the
glass be always heated in it: and so it may be cut.   In
the same manner may other stones"——

His directions for cutting crystal are so unintelligible
to me that it will be more advisable to give them in the
original than to attempt a translation, at the risk of substi-
tuting my own ideas for his:—

"*QUOMODO CRISTALLUM POSSIT SECARI.*

"Cristallum facile tali valet arte secari :
    Opportuna tibi quæratur lamina ferri,
    Huic etiam binæ claves jungantur utrimque
    Ex ferro, medium quæ firmant undique plumbum ;
    Nam plumbo solo tribuetur cura secandi,
    Ipsæ custodes laminæ sint exteriores
    Ut tibi dent rectum recto conamine cursum.
    Sed nec duritiem poteris prorumpere tantam
    Mollitie plumbi nec (nisi ?) quiddam junxeris illi—
    Tanquam pulverulas fornacis fragmine micas
    Contere, quas teneræ poteris connectere lamnæ ;
    Hæc etenim plumbum conjunctio reddit acutum
    Et quum (suum) rursus habent lateris fragmenta vigorem—
    Sed vim cristalli cruor antea temperet hirci."

The sense seems this : to strengthen a plate of lead with
two parallel bars of iron, which may serve as a guide to
the hand in rubbing the crystal; and to give the lead

a cutting surface, sprinkle it with powdered potsherds,* which get fixed in it by the pressure of the crystal upon them, just as in making watch-jewels at present the coarse diamond-powder, the *Bort*, is fixed upon the surface of a soft-copper disk by a few blows of a hammer before it is set a-going.

PASTE GEMS.

Although the Roman imitations of the precious stones— in two at least, the Emerald and the Lapis-lazuli—were truly admirable, yet in all the other kinds, except as to hardness, they are far outdone by the modern Parisians. I transcribe from Barbot's 'Traité des Pierres Précieuses' (*Strass*) the ingredients and proportions for making the most important species :—

Strass.

Rock-crystal	220·07
Minium	342·17
Potass, purified	116·90
Borax	15·07
Arsenic	0·66

Emerald.

Strass	250·00
Copper, green oxide ..	2·51
Chrome, oxide	0·11

Beryl.

Strass	187·50
Antimony, glass	1·32
Cobalt, oxide	0·082

Topaz.

Ceruse of Clichy	50·0
Quartz pebbles calcined.	50·0

Sapphire.

White strass	31·25
Cobalt oxide	0·11

Another.

Strass	500·00
Copper, acetate	5·9
Iron, tetroxide	0·825

Garnet, Siriam.

Strass	27·75
Antimony, glass	13·95
Purple of Cassius	0·110
Manganese oxide	0·110

---

* Perhaps meaning the same substance as Pliny's "ostracias, testacea durior," used for cutting other stones (OSTRACIAS).

Strass, so called from the German chemist who invented it, is the brilliant White Paste, the material of Paste Diamonds. It will be seen that, coloured by different metallic oxides, it reproduces the other gems. Barbot gives many varying recipes for its composition ; but they have, as far as the Diamond is concerned, been superseded by a quite recent discovery of the ingenious Parisian chemists. By a preparation of borax (as yet a secret) a paste, the " diamant de borc," is produced as hard as quartz, defying the file, and when set in real gold (as such now are) not to be distinguished by the most practised eye from true brilliants.

ENAMELS.

Enamels, though the same in composition as pastes, differ in the mode of application. The vitreous mass is ground fine, and then, mixed with gum-water, is applied with a brush to the surface to be ornamented, upon which it is finally fixed by means of fusion. From the degree of heat necessary in the operation, the substratum must be either fine gold or pure copper, capable of resisting it, otherwise the slight compartments formed in the metal to contain the enamel would run, and the outlines of the pattern be deranged. The basis of all ancient enamels, for which Mediæval writers give many recipes, is powdered glass or flint, oxide of lead, and borax, mixed in various proportions, and coloured by different metallic oxides. It was a Celtic invention, made to replace the Λιθοκόλλητα, or inlaid gem-work of the Orientals, by a cheap and attainable imitation in the baser materials of glass and copper. This we learn from a passage in that pretty piece of Philostratus, the *Imagines,* or 'Picture Gallery,' where he is describing the Boar-hunt (xxvii.) :—" The horses have their bits plated with gold, and their head-ornaments are likewise gilt and diversified with colours. These colours, it is said, the barbarous nations (of the West) dwelling upon the ocean pour into the copper when red-hot, so that they become like a stoné when congealed, and retain the position traced out for them." An ambiguous account of the process, which shows how little was known about it in Rome at the time he wrote, the reign of Severus. But in Britain the art had been carried to perfection long before this, as is attested by the remarkable incense-burner—its surface covered with a tasteful floriation in red and blue—found with the other relics (a curule-chair and bronze vases of Greek workmanship) in the tomb-vault of one of the Bartlow Hills. Numbers of fibulæ, similarly decorated and of British origin, are to be seen in collections. All these are done by the *champ-levé* process: that is, the patterns have been cut out to a con-siderable depth in the metal, and these beds filled up with the fused enamel, afterwards polished down to a smooth face. Such also was the mode in use with the Gothic artificers throughout the Middle Ages, as the innumerable reliquaries, coffrets, and tablets of French and German manu-facture, sufficiently testify. But the Byzantine jewellers took it up and applied it to gold, inventing the *cloisonné* method, more exactly imitating the old λιθοκόλλητα, each colour being contained in its distinct compartment of thin gold-plate set on edge upon and soldered down to a stouter basis, the thin lines of the gold serving for the outlines of the whole design, which is often a singularly complicated piece. This Byzantine art was adopted by the Persians in the East, as the wonderful cup of Chosroes (to be de-scribed hereafter), in the Bibliothèque Impériale, remains to declare, and by the Anglo-Saxons in the West. The very elaborate jewel made by order of King Alfred mani-

fests their proficiency in the art; and several other examples
of Saxon skill in cloisonné enamelling ornament our collections. This peculiar method yet lingers in Algeria; a
tradition of Byzantine dominion. Kabyle-made bracelets
were lately shown me, adorned at intervals with square
plaques in silver encrusted with cloissoné enamels so
purely Byzantine both in pattern and in technique that
had they been found in this soil they would infallibly have
been put down as importations from Constantinople prior
to the Norman Conquest.

The third and modern manner of enamelling upon a
smooth surface came later into use, not before the middle
of the 15th century, being merely an adaptation to copper of
the Moresco invention of painting with enamel colours upon
a ground of stanniferous glaze laid over earthenware, which
the Italians of those times were carrying out with so much
success in their majolica. In this last development of the
process the vitreous glaze was merely laid on like water-
colours upon the polished copper, with no longer any bed
traced to direct it, the dexterous application of the fire alone
preventing the several colours from flowing, when fused,
out of their proper positions. Of this style the best types
are the pieces of Limoges work, now so much sought after,
specimens of which appear soon after 1450, that is contempo-
raneously with the beginning of the majolica manufacture
in Italy; but the new method only attained perfection in
the middle of the next century, when the paintings of the
Rémonds display a manner as fine as that of their Italian
rivals in the other branch of glaze ornamentation. Some
nicely-finished paintings in this style may be found adorning
jewels of the Cinque-cento; but it was not before the age
of Louis XIV. that the French, the great masters in this
art, headed by the inimitable Petitot, produced miniatures

on gold as exquisitely finished and as spirited as the best paintings upon ivory.

This also is the method employed by the Chinese. It seems at variance with the nature of that race to suppose they learnt the art from the Europeans. In fact they have applied it to works on a much larger scale than were ever attempted elsewhere. We have from that country tall ewers, capacious basins, elaborate vases, perfectly painted with enamels of the most vivid colours, and possessing this superiority over European enamelled work, that they are not injured by the contact of boiling water.

CUP OF CHOSROES I.

The sole relic of the splendour of the greatest of the Sassanidæ, the "Nourshirwan the Just," so famed in Persian story, is the Cup of Chosroes I., now deposited in the Bibliothèque Impériale, Paris. For more than ten centuries had it been preserved in the treasury of the Abbey of St. Denys, to which it had been presented by Charles the Bald; where it was ever held in the highest veneration as the very cup of King Solomon, "whose figure (according to Dom Doublet), drawn after the life, seated upon his throne, with steps adorned with lions on each side, as Holy Scripture represents him, appears cut in relief upon a very excellent and large white Sapphire set in the bottom. Also is it enriched around the brim with Jacinths, and inside with very fine Garnets and very fine Emeralds." Thus its value as a Scriptural relic was enhanced by the supposed inestimable worth of the precious stones enclosed within its circumference. But, alas! these gems, subjected to the severity of modern criticism, prove to be but paste and crystal, and the ownership has been

transferred from Solomon to a monarch later by fifteen
centuries, though equally renowned for his wisdom amongst
the Orientals. In the signal defeat of the ambitious Shah
by Justinian, the general of Tiberius Constantine, on the
plain of Melitene, when "a Scythian chief who commanded
their (Roman) right wing, suddenly turned the flank of
the enemy, attacked their rear-guard in the presence of
Chosroes, penetrated to the midst of the camp, pillaged the
royal tent, profaned the eternal fire, loaded a train of
camels with the spoils of Asia, cut his way through the
Persian host, and returned with songs of victory to his
friends" (Gibbon, ch. xlvi.), this, the drinking-cup of the
king of kings himself (the sacred κόνδυ), may with the
greatest probability be assumed to have formed a part,
and the most signal trophy of the Roman success. In the
constant interchange of friendly offices between the Byzantine and Frankish courts, it may easily have found its way
to France within the ensuing two centuries, before the time
when Charles the Bald appears as its possessor. To abridge
the accurate description, by Chabouillet, of this unique
specimen of the art of the goldsmith, glass-worker, and
gem-engraver of the 6th century : " This transparent cup is
composed of a sort of framework in massy gold, and of
three circular rows of eighteen medallions in rock-crystal
and coloured glass,* serving as a frame to the principal
subject, a round medallion of rock-crystal, filling the
bottom or 'umbril' of the bowl. This is engraved in
relief, but set with the plane surface inwards, and repre-

---

* In imitation of the more precious materials similarly employed in
goblets of the Early Empire. "Nor did these suffice (the Murrhine
and Crystal), we drink out of an assemblage of gems, and weave cups
out of emeralds, and exult to hold India in our hand for the purpose of
intoxication, whilst gold is a mere accessary," which last evidently
alludes to the mere skeleton gold frame-work holding the gems together.

sents Chosroes I. seated upon a throne, the supports of
which are winged horses. At the back of the throne is a
large cushion, but thrown to the right by an error in the
perspective, not to be wondered at in a work of this date.
The monarch is crowned with a globular turban, surrounded with battlements and surmounted by a crescent,
over which appears a second crescent supporting an orb,
whence depend two streamers. By the form of the crown
we identify this portrait as designed for Chosroes I., for it
exactly coincides with that worn by him upon his gold
medal in the Blacas Cabinet; and it is a well-known rule
that each Sassanian king modified the form of his head-
dress, and the attributes decorating this venerated mark of
sovereignty, upon his coinage. The monarch is shown in
front-face; his hair, divided in two huge curly clusters,
falls upon his shoulders; his beard, thick but short, and
not curled. He is clothed in an embroidered robe or *candys*,
and rests both his hands upon the pommel of his sword in
its scabbard. From his shoulders float two large streamers,
like those of the turban, and two still larger from his bust,
spreading away towards the left, and two more from his
feet. Both the border of the bowl and the mounting of the
large medallion are inlaid with cubes of glass imitating
Garnets. The medallions of the rows above mentioned are
alternately white and violet: the former in rock-crystal,
engraved on their outsides with a flower in relief; the latter
in glass, cast with a similar ornament. The interstices
between the medallions are filled up with lozenge-shaped
pieces of glass of a transparent green. Dimensions of the
bowl, 28 cent. wide by 35 deep (or about 11½ inches by
14)."

The traditional celebrity of this splendid goblet, the
pride of the most illustrious of their ancient sovereigns,
may well be supposed to have furnished Oriental story-

tellers with the legend of King Jamshid's famed Ruby Cup, of which Hafiz sings :—

> " My wanderings brought me to the fast-closed niche
> Where Jamshid's cup of sculptured Ruby lay;
> Blushed through the porphyry wall its radiance rich,
> As through the curtains peeps the rising day.
>
> Earth and earth's gifts were graven on its round,
> Her cities, nations, of all tongues and kin,
> The various treasures in her bosom found :
> And words of power the buried wealth to win."

Note on the Portland Vase.

That this work is due to the Ptolemaic school at its best time is deducible not merely from the purity of the style, but yet more from the recondite character of the designs and their enigmatical expression of some well-known myth, all bespeaking the over-refined genius of the age when flourished Callimachus and Lycophron. Roman compositions, on the other hand, always tell their own story under the thinnest possible mythological disguise. If the courtship of Peleus and Thetis be really intended by the one relievo, her bereaval of her son by the other, the vase may reasonably be supposed a commission from some prince of the Macedonian line, like Perseus,—

> " proavi simulantem pectus Achillis."

That its latest *ancient* owner—the Roman whose ashes it accompanied (not *contained*)—was the virtuous but ill-starred Syrian, must be a mere *antiquario's* fiction. That emperor was murdered, with his mother and all his friends, in some remote part of Gaul, or perhaps Britain (Lampridius is uncertain which); and his successor, the instigator of the crime, was the last person to honour his remains by transmitting them to Rome for imperial sepulture.

DESCRIPTION OF THE WOODCUTS.

Titlepage. No gem could be better chosen than this to give a soul to the titlepage of a book treating upon the special materials of the Glyptic Art; bearing as it does the most ancient as well as the most complete of the few *genuine* artists' signatures known; for it proclaims itself " Engraved by Dexamenus the Chian." Its high antiquity is demonstrated by the fact of the letters reading from right to left in the impression (the way in which they were intended to meet the public eye), which proves the work prior at least to the date of Herodotus (B.C. 450), for he speaks (II. 36) of the present mode of writing Greek as the regular one in his times.

In the choice of the type—the flying Stork—it is more than probable the engraver intended to express by a *rebus* his own nationality; his belonging to that ancient race, the Pelasgi, whose name (" Storks ") their Hellenic rivals deduced from their frequent, though enforced, migrations. Of this same race, Mnesarchus, the earliest gem-engraver on record, is known to have been, for he was a native of one of the isles in the Ægean out of which the Athenians expelled the Tyrrheni, whereupon he sought an asylum in Samos.

Engraved on the base of a sard scarabeus of exquisite workmanship found at Kertch; and now in the Russian Cabinet.

P. viii. Apollo reposing after his victory over Marsyas. He rests his lyre upon the head of Themis, the original patroness of Delphi, who holds forth a branch of the *fagus*—chestnut-tree—whereof was woven the garland, prize in the Delphic Games. Sard; slightly enlarged.

P. 13. Queen Elizabeth's Agate, described at p. 10.

P. 21. The Sassanian imperial standard : a favourite seal-device under that monarchy. Its meaning is placed out of doubt by the legend "Afzud direfesh," "Long live the royal banner," frequently accompanying it. The symbol appears to denote the sun and moon conjoined and placed upon a tripod ; the same being also an ancient Buddhist device. Pehlevi legend, not yet interpreted. Almandine.

P. 26. The " Golden E " of Delphi ; the sacred number Five : also explained as implying εἶ, " Thou art," addressed to the Deity. Originally a Hindoo caste-mark (of the votaries of Vishnu), but placed in another position so as to admit of a Grecian significance. Agate-onyx, cameo.

P. 36. Silenus ivy-crowned. Sard.

P. 39. Dolphin and trident : the letters in the field probably refer to the owner of the signet. Sard.

P. 42. Mithraic genius, holding the sacred asp, and *situla* of holy water : addressed in the legend on the reverse as the god Phren, the Egyptian title of the Sun. Green Jasper.

.P. 49. Amulet against the Evil-eye : representing that dreaded object surrounded by the attributes of the deities to whom the *seven* days of the week are consecrated ; whose protection is thus secured to the wearer. An interesting memorial of the present naming of the week-days being already established in the second century, the date of this intaglio. Sard.

P. 69. Sassanian princess : the crescent in the field

declares her rank. Legend " Rozozi," " Lady of the Spring."
Red and white Agate.

P. 76. Scipio the Elder, wearing the *Cudo* leather skull-
cap, to conceal his baldness. Sard.

P. 81. Marine Cupids: one bestriding a dolphin, the
other holds up the tail of a second fish to catch the breeze.
Red Jasper.

P. 92. Proserpine: perfect Greek style. Sardoine.

P. 99. Capricorn conjoined with Scorpio, carrying a
legionary standard. The former being the badge of the
second Augustan Legion, the latter the *tutela* of Mars ; it
may reasonably be inferred that this was the seal-device of
some officer in that corps ; who thus courted the favour of
the general and particular patrons of the service. Sardoine.

P. 103. Marine genius, wielding the trident of Neptune, and mounted on " tyrannus—Hesperiæ Capricornus
undæ," whose influence must have been specially felt by the
coral-fishers. Sard.

P. 122. Cupid peering into a huge crater whence escapes
a spectre-skeleton ; an enigmatical device, perhaps typical
of the antagonism of the Principles of Nature. Onyx.

P. 126. Drusus Junior: a work in the finest Roman
style, and worthy of Dioscorides, to whose hand it may
with some reason be attributed. Sardoine.

P. 129. Horus, the sun-god, seated on the lotus, emblem
of fecundity ; adored by the cynocephalus, the attribute of
the moon. In the field are seen the solar star, and triangle,
symbols of both luminaries. Green Jasper. (Waterton.)

P. 132. Julia, daughter of Augustus, in the character
of Isis. Sardoine.

P. 136. Victory in the car of *Ceres*, probably allusive
to the restoration of Africa to the Empire. Sard.

P. 151. Victorious discobolus, holding the discus and
the wand, *calamus*, ornamented with ribbons, in use for

measuring the cast, like the old *reed* in the game of bowls. The prize-vase of oil stands on the table. Plasma of the purest quality.

P. 168.   Elephant's head: Pehlevi legend, " Masdaki Raj . . .," " Prince Masdaces :" probably some Indian tributary; not the communist philosopher, the adoption of whose doctrines cost King Cavades his throne.   Cinnamon-stone.

P. 191.   Faustina, as Ceres, in a *theusa* drawn by elephants: symbolising, as the legend on her coin with the same type expresses it, the idea of eternity.   Yellow Sard.

P. 202.   Astrological gryllus composed of a Silenus-mask, and heads of a ram and horse, the*attributes of Mercury and Neptune.   The sun and moon in the field indicate the true purport of this talismanic device.   Sard.

P. 208.   Actor wrapped in the *pallium* wearing a mask with gaping mouth, and holding the *pedum*, distinctive badge of Comedy.   Greek work.   Sardoine.

P. 212.   Seated sphinx; the type upon the coinage of Chios.   Greek work.   Black Jasper.

P. 238.   Mars Gradivus carrying a trophy: of the best Greek period.   Onyx of three shades cut athwart the strata.

P. 247.   Syren with lyre, an attribute which coupled with the inscription KAΠ, and the palm, indicate the signet of a successful competitor for the poetic prize at the *Capitoline* Games.   Sard.

P. 258.   The child Opheltes destroyed by the Lernæan serpent: to commemorate whose fate the Isthmian Games were instituted.   Middle Roman period.   Red Jasper.

P. 260.   Plastes modelling a bust in wax or clay supported on a *turbo*, cylindrical stand, made to revolve at pleasure.   This contrivance was in use also at the slave-market for exhibiting the *points* of handsome youths to purchasers.   Statius says of a home-raised favourite,

" Non te barbaricæ versabat turbo catastæ."

P. 268. The Fortuna *Fortis* of Antium, as she is figure with her twin-sister, *Felix*, on a denarius of the *Rustia* Gens : on whose family name the rural ant in the field is the rebus. Prase.

P. 272. Cybele, or, rather, Faustina Mater, thus deified. Red Jasper.

P. 277. The portrait noticed at p. 275. It has very much the character of that of Demetrius Poliorcetes on his medals. The horn is clearly not the wreathed one of Ammon, but rather the wild goat's of Caranus, founder of the Macedonian dynasty : assumed by the present bearer on the same principle as Alexander did that of his parent-god, and Lysimachus that of Bacchus.

P. 286. Young Faun acting the cup-bearer to his senior, and filling his scyphus with his œnochoe out of a crater proportionate to the thirsty nature of the family. Archaic Greek. Yellow Sard.

P. 299. Achilles about to arm himself; in one hand he holds his sword, with the other he draws forward his chlamys ; the several pieces of the panoply lie around him in readiness on the ground. A very minute engraving. Sardonyx.

P. 302. Sol, lord of the universe, in his four-horse car. Around the field the Gnostic invocation ABΛANAΘA-NAΛBA, " Thou art our Father ! " In the exergue, TYΞEYI, which Alex. Trallianus informs us is a title of the Sun. Loadstone.

P. 311. The Cæsar Diadumenian (p. 308) as Mars : but wielding, instead of the proper spear, the rudder of Fortune : a flattering augury, disproved by his untimely and miserable fate. Sard.

P. 322. Recumbent Sphinx : archaic Greek work, retaining much of the Assyrian manner. Scarabeoid : Sardoine.

(G.)   2 B

Pages 370, 377.　Obverse and reverse of a seal-tablet in fine yellow Jasper, bearing the name, in a cartouche, of Amenophis II., whose date is the 15th century before our era.　The extremely free and natural drawing of the bull, however, proves of itself that this intaglio cannot belong to that remote period, or indeed to pure Egyptian art of any period, but that it is either a Phœnician or even a Ptolemaic copy of an earlier and ruder original.　The peculiar type of horse, observable on certain Phœnician medals also, makes me incline to the former explanation of the parentage of this exceptionally excellent specimen of the Egyptian style.　(British Museum.)

INDEX.

ABPAΣAΞ.

ABPAΣAΞ, IAΩ, 36.
Abyssinian trade, ancient, 295.
Achates, Drillo, 1.
——, "fidus," how explained, 7.
Adadu-nephros, 207.
Adamas, the first, 201.
Adder-bead, Druid's-bead, 251.
Adularia, 300.
Æstyi, amber-gatherers, 306.
Aëtites, ordeal by, 12.
——, medicinal virtue of, 13.
Agate, where found, 1, 87.
—— distinguished from Onyx, 2.
——, virtues of, 6.
—— of Pyrrhus, 7.
—— of Queen Elizabeth, 10.
——, false, 342.
Agathe-Onyx, 222.
Agricola, 257.
Alabaster, Onyx-marble, 22.
——, origin of name, 23.
—— used for Camei, 26.
Albâtre-Onyx, used for Italian Camei, 222.
Alexandria, seat of crystal-working, 107.
Alexandria, glass fabrique of, 236, 337.
Almandine garnet, 16.
Amazon-stone, highly valued by the Aztecs, 127.
——, antique works in, 128.
Amber, malleable, 310, 352.
——, largest piece known, 306.
—— used for false gems, 307.
—— containing insects, 309.
——, medical virtues of, 310.
—— used in ancient glyptics, 303.

BOHEMIAN.

Amber, confounded with jacinth, 304.
—— origin, accounts of, 305.
Amethyst, common, 27, 139.
——, oriental, ib.
——, origin of name, 28.
——, virtue of, 36.
Amethystus, the ancient, 32.
——, described, 35.
Amulets of coral, 102.
Anæsthetics, ancient use of, 256.
Anthracitis, 272.
Ἄνθραξ of Theophrastus, 14.
Apanturin, brocatella, 270.
Arsinoe in crystal, 115.
—— in peridot, 313.
Assaying of silver, Roman, 154.
Aster, Astroites, Astrobolus, 39.
Asteria, star sapphire, 37.
"Audaces calices," 334.
Auldjo Vase, the, 331.
Aurichalcum, 158.
Aurungzeb, topaz of, 96.
Aventurine, its composition, 269.
"Axinomantia," divination by jet, 130.

Babylonian Sard, 282.
"Bacco, tomba di," 263.
Bacchylides. 152.
Barygaza, Baroche, 227.
Basalt used for statues, 40.
Basanites, basalt, 42.
Batne, fair at, 295.
Batrachites, mottled jasper, 43.
—— toadstone, 44.
Bohemian garnet, 17.

2 B 2

BEAUTY.

Beauty, Roman idea of, 308.
Belemnite, a Celtic jewel, 81.
Beli oculus, 321.
Belus, river, 349.
Beryl, the ancient, 50.
——, falsified, 52.
—— of Julia Titi, 53, 337.
——, origin of "brille," 55.
"Beryl" for Sard, 282.
Bitumen used in vase painting, 130.
Black amber, jet, 131.
Blacas Cabinet, 31, 36, 275, 293, 318, 322.
Black jasper, and trachyte confounded with obsidian, 210.
Bloodstone, why so named, 136.
—— used for religious camei, *ib.*
Boar, emblem of Hertha, 306.
Bonus Eventus, 277.
Borax diamond, 358.
Brunswick Vase, the, 225.
Bunwary, Hindoo touchstone, 154.
"Burning of Troy, the," 241.
"Burnt Sard," imitative Sardonyx, 285.
Byzantine Camei, 276.
—— enamels, 359.

"Cabochon" cutting, 17.
Cadmea, Cadmia, 248.
Calcedony, derived, 82.
—— used for phaleræ, 84.
——, the ancient Achates, 87.
—— commonly used for Cinquecento works, *ib.*
"Calf's-eye," the, 245.
Callaina, where found, 57.
——, the peridot, 58.
Callais, 59.
Calligone's Sardonyx, 288.
Callimus, eagle-stone, 11.
Cairn-gorum, Rauch-topaz, 94.
"Cameo," etymology of, 70.
—— in shell, 73.
—— false, 222, 237.
Camei used for talismans, 72.
—— mistaken for nature-pictures, 9.
"Camée, le grand," 292.

COPPER.

Camillo di Leonardo, 43.
Capnias, 86.
Caranus, wedding of, 157.
Caravan-trade in Africa, 295.
Carbuncle, the modern, 17.
Carlovingian intagli in crystal, 117.
Carnelian, derived, 278.
——, virtue of, 286.
Carpegna Cameo, the, 291.
Canopic vases, 23.
Capis, its figure, 185.
Carnuntum, a Celtic entrepôt, 306.
Castel-Bolognese, crystals of, 116.
Cat's-eye, common, 319.
——, chrysoberyl, 320.
——, lion's head in, 320.
Caylus, onyx vase of, 226.
Cayman-stone, 45.
Cellini's recipe for gold-solder, 89.
Celts, Greek trade with, 305.
Cerachates, white Calcedony, 4.
Ceraunia, ruby, 77.
—— opal, 242.
—— stone-celt, 78.
Chalcedonian emerald, 82.
Chardin visits the Shah's treasure-house, 64.
Chastity, test of, 130, 173.
Chinese mixed-metal vases, 157.
Chinese agates, artificial, 10.
Choaspes, agates of the, 1.
Christopher, St., influence of, 270.
Chosroes, cup of, 361.
Chrysoberyl, Indian chrysolite, 51.
Chrysocolla, gold-solder, 88.
Chrysolite, chrysoberyl, 99.
Chrysolithus, and its varieties, 93.
Chrysoprase of the moderns, 90.
Cinnamon-stone, garnet, 17.
——, chryselectrus, 166.
Citrinus, 168.
Claudius, porphyry statues of, 261.
Claudian on the Enhydros, 119.
Cleopatra, signet of, 28.
"Cloisonné" enamel, 357.
Cochlides, lumachella, 233.
Constantia, tomb of, 262.
Constantius II., cameo, 293.
"Copper, precious as gold," 157.

CORAL.

Coral, ancient idea of its nature, 100.
—— much used by the Gauls, 101.
——, artificially blanched, 103.
Corallis, red jasper, 146.
Coralachates, 7.
Corundum, 201.
Cosimo I., carves porphyry, 264.
—— and family, cameo, 294.
Cosmus, 227.
"Corinthian Brass," the, 155.
"Cotes," emery-stones, 190.
——, Armenian, 191.
Creta, sealing, 163.
—— matrix, 339.
Crystal, the Ethiopic, 104.
—— used for vases, 105.
—— for a burning-glass, 110.
—— in glass-making, 112.
—— for false gems, 113.
—— for statuettes, 115.
—— to soften, 355.
Cymophane, 38, 320.
Cyanus, not the sapphire, 123.
——, a pigment, 125.
——, artificial, 277.
Cymbia, gondoles, in chrysoprase, 90.
Cynthia's beryl, 51.
Czernovitza opal mines, 241.

Dante, 134.
"David," gem of king, 337.
Dee, Dr., showstone of, 122.
"Devil's looking-glass," Kelly's, 209.
Diadochus, 122.
Diamonds, parturition of, 12.
Diana de Poitiers, bracelets of, 74.
Didius Julian cameo, 293.
Doublets, 114, 334.
Dracontia, the diamond, 48.
Dresden Sardonyx, the, 298.
Duchanek, lapis spiritalis, 252.

Earinus, pyxis of, 291.
Egyptian pebble, 5.

GOA-STONE.

"Egyptian-stone," basalt, 40.
Egyptian mosaic, 328.
Elizabeth, Queen, agate of, 10.
Electrum, 310.
Emery, 192.
——, where found, 195.
Emerald, root of, 265.
Enamel, invention of, 358.
Enhydros, 119.
Essonite garnet, 17.
Etruscan jewelry, 89.
—— pastes, 327.
Eye-onyx, 321.
——, singular adaptation of, 322.

Farnese tazza, 225.
Felspar, 127.
Firmus, Indian trader, 228.
——, glass palace of, 337.
Florentine pietra-dura mosaic, 151.
Foils, how made, 159.
Fortuna Seia, 25.

Galen on the virtue of the green jasper, 149.
"Garnet," derivation of, 14.
——, Siriam, 16.
——, engraved, 18.
——, vases of, 19.
Gem-engraving, origin of, 195.
—— of the Aztecs, 198.
—— of the Jews, 201.
Gems found in animals, 45.
——, whence obtained by the early Greeks, 234.
German onyx, how stained, 232.
Girasol, semi-opal, 244.
Glass, malleable, 351.
——, to engrave, 354.
——, British, 349.
—— first noticed, 325.
—— antique vases, 332.
—— medallions, 335.
—— coin, 349.
——, Venetian, fabric of, 350.
Goa-stone, 256.

GOAT'S-BLOOD.

Goat's blood, a solvent for gems, 354.
Guaruaccino, red jacinth, 164.

Hadrian, basalt statue of, 40.
Hæmatinon glass, 329.
Hæmachates, 3.
Hæmatites, Hæmatitis, distinguished, 146.
Heliodorus, 33.
Helen, cup of, 311.
——, talisman of, 38.
Heliogabalus, 186.
Heliotrope and bloodstone, 133.
——, magic use of, 135, 279.
Heraclius, "De Artibus Romanorum," 201, 339, 355.
Helena, tomb of, 262.
Honey, used in treating gems, 232.
Hope sun-opal, the, 320.
Horace, portrait of, 318.
Hydaspes, ring of king, 259.
Hysteropetra, 208.

Icterias, jaundice-stone, 168.
"Imagines" of Philostratus, 358.
Indian glass, porcelain? 111.
—— exports to Rome, 227.
Iris, artificially made, 80.
"Ivory, the Fossil," of Theophrastus, 60.

Jacinth, 94.
——, engravings in, 162.
——, origin of name, 167.
——, a cure for jaundice, 168.
Jade, property of, 203.
—— carved by the Aztecs, 204.
—— by the Chinese, 206.
—— known to the Celtic Helvetii, 205.
Jasp-onyx, 142.
Jasper of the moderns, 144.
Jaspis of the Greeks, 137.
—— of the Romans, 141.
—— used for phaleræ, 143.

MALACHITE.

Jesù, chiesa del, altar front of, 191.
Jet ornaments, antique, 132.
—— anciently confounded with bitumen, 129.

Kabyle enamels, 360.
Kaman, Kakaman, Camahen, source of "cameo," 72.
Kertch, tomb-treasures of, 165.
Kimmeridge-coal, 131.
"Kings of Cologne, the," 74.

Labrador-spar, mithrax, 129.
Lapis-lazuli, 273.
——, Egyptian, 274.
——, works in, 275.
—— paste, 277, 330.
Latrunculi, 337.
Leopard's head, Townley, 221.
Leo X., seal of, 264.
"Life, fountain of," 316.
Ligurius, 161.
—— explained, 304.
Loadstone statue of Venus, 174.
Louis XIII., in opal, 247.
"Loupe," derived, 55.
"Lupa," rough sapphire, 55.
Lumachella, 271.
Lycophthalmas, 321.
Lydius lapis, 152.
Lyncurium, origin of, 160.
——, the jacinth, 162.
Lynx-tree producing amber, 305.

Magnet, varieties of, 169.
—— used for talismans, 171.
——, magical virtues of, 172.
——, an ingredient in glass-making, 173.
Manganese, Magnes, 174, 350.
Matrix of Opal, 271.
Malwa, "land of Havilah," onyx mines of, 228.
Malachite, an amulet, 177.
——, the Smaragdus Medicus, 178.

MARBLES.

Marbles, antique, whence obtained, 234.
Marlborough Gems, 97, 142, 276, 285, 293, 301.
Matidia in calcedony, 86.
Magnifying-glass, first invention of, 55.
Maria Honorii, tomb of, 105.
Marculus, cælum, 283.
Marcasite, 250.
Mausolus, Sardonyx of, 299.
Mexican inlaid work, 62.
Mirror, divination by, 122.
——, ancient material of, 350.
Mithrax, 129.
Mithridates, vases of, 184.
Morio, Mormorio, dark jacinth, 164.
Molochites, green jasper, 176.
Mosaic, invention of, 336.
Mugnone, 279.
" Mulc " Calcedony, 86.
Murrhina, described, 179.
——, supposed nature of, 181.
——, extant specimens of, 183.
——, ancient value of, 185.
——, porcelain so called, 189.
Murrhine glass, 331.

Napoleon, tomb of, 263.
Nature-pictures in agate, 5, 9.
Nechepsos, 150.
Nephrite, jade, 203.
Nero, jasper statuette of, 86.
——, murrhina of, 185.
—— promotes the amber trade, 306.
Nero antico, 210.
New Zealanders, jade carved by the, 205.
Nicolo, 220.
—— unknown to Greek art, *ib.*
——, why valued by Jews, 221.
Nile, old bed of the, 1.
Niger Pescennias, basalt statue of,41.
Nsao, crapondinus, toad-stone, 43.

Obsidian mirrors, 209.
——, the Nero Antico, 210.

PERSEUS.

Obsidian, Aztec weapons of, 212.
Oculus Mundi, hydrophane, 255.
Olivine, Callaina, 68.
——, found in aërolites, 69.
Omphax of Theophrastus, the chrysoprase, 91.
Opal, cause of its iridescence, 239.
—— of Nonius, 240.
—— of Josephine, 241.
——, influence of, 246.
Ophites, the oracular, 254.
——, the Salagrama, 255.
Opus Alexandrinum, 261.
Onyx, derived, 213.
—— first mentioned, 214.
——, its nature defined, 217.
Onyx, imitative, 222, 331.
——, vases of, 224.
——, mines of, 227.
——, how improved by the Hindoos, 230.
——, by the Germans, 232.
—— imitated in glass, 235.
—— in composition, 237.
Onychitis, vases of, 184.
Ostracias used in gem engraving, 248.
—— perhaps pyrites, 249.
Ovum anguinum, 251.
Owl in lapis-lazuli, 275.

Pæderos, amethyst, 29.
——, opal, 243.
Pantarbes, the ruby, 259.
Pallidus, χλωρὸς defined, 57
Pan-fish, and gem, 38.
Pastes, how made, 323.
——, antique, hardness of, 325.
——, Egyptian, 326.
——, Roman, 329.
——, Renaissance, 342.
—— precious stones, 358.
Paste camei, 330.
Peridot first discovered, 312.
——, statuette in, 313.
——, name derived, 317.
Perfumes, ancient value of, 23.
Perseus, signet of, 275.

PETITOT.

Petitot, 360.
Phaleræ, 85.
Phengites used for glazing windows, 23.
Phryne, intaglio, 282.
Philip II., in topaz, 97.
Plythanæ, Pultanah, 227.
Porcelain, first imported, 188.
——, early notice of, 111.
Polemo, epigram of, 137.
Porphyry, quarries of, 261.
——, how cut, 264.
Pollexfen Vase, the, 332.
Polycrates, ring of, 298.
Portland Vase, the, 331, 364.
Plasma, derived, 265.
Prase, plasma, intagli in, 267.
Prehnite, 266.
Ptolemies, cup of the, 224.
Pyrites used in gem-cutting, 249.
—— as marcasites, 250.
Pyrope, Bohemian garnet, 19.
Pyropus, 158.
Pyrrhus, Agate of, 7.

Quartz, amethystine, 27, 139.
Queen of Sheba's portrait, 337.
—— Syra's necklace, 303.

Red jasper only used by the Romans, 145.
Regent's pastes, the, 343.
"Repentance, stone of," 316.
Rings of crystal and calcedony, 109.
Rothschild Vase, the, 333.
Rubace, "ruby of Ancona," 113.
Rubies, imitative, 20.

Salagrama, 255.
Samir, the worm, 200.
Sandaster explained, 271.
Sapphirine calcedony, 84.
Sapphirus of the ancients, 273.
Sarcophagus in bloodstone, 136.
Sard, derived, 278.
——, where found, 279.

THUNDER-STONES.

Sard, special merit of, 280.
——, varieties of, 281.
Sardo, Mount, 279.
Sardoine explained, 282.
Sardonyx, defined, 287.
——, why perforated, 290.
——, largest known, 291.
——, camei in, 291.
——, Byzantine, 294.
——, imitative, 284.
—— in paste, 296, 330.
——, present value of, 297.
Sardonyx, test of genuine, 232.
Schmelze glass, 337.
Scorpion on jasper, 150.
Selenites, moon-stone, 300.
Sealing-wax, antique, 163.
Serpentine, 257.
—— used by the Assyrians, 258.
Seleucus, cup of, 326.
Shell-cameo, 73.
Sibyl, legend of the, 163.
Sicilian agates, 1.
Signatures, the doctrine of, 168.
Sinai, turquois of Mount, 64.
Sirius, the Marlborough, 18.
Siriam garnet, 16.
Smaragdi, monster explained, 128.
Solis gemma, 300.
"Solomon, the cup of," 361.
Sphragides, yellow sards, 140, 281.
Sun-stone, 301.
Staticula, ροῖσκοὶ, 4.
"Strahl-pfeilen, strahl-hammern," thunderbolts, 75.
Strass, 357.
Styx, water of, 183.
Stone celts, how made, 198.

Tables of the Law, material of, 274.
Tacitus, an amateur in glass, 333.
Talc, 302.
Tagarah, Dowlatabat, 227.
Talismans on amethyst, 35.
"Tallow-drop" cut, 17.
Tanos, Amazon-stone, 128.
Tassie's pastes, 346.
Thunder-stones, 81.

TIBERIUS.

Tiberius, bust in turquois, 63.
—— in calcedony, 86.
Topaz, derived, 312.
——, Brazilian, 316.
——, pink, how made, 317.
——, white, blue, 318.
——, oriental, 93.
—— of Aurungzeb, 96.
Toreuma, cameo, 332.
Touch-stone, ancient, 44, 152.
——, how used by the Italians, 153.
Tree-agate, 6.
Tricoloured, or banded, agate, 2, 214.
Trulla, murrhine, 183.
Turquois, 61.
—— used by the Mexicans, 62.
——, Persian, inlaid, 63.
—— "de la vieille roche," 64.
——, virtues of, 65.
Tutenague, known to the Greeks, 155.

Ultramarine, "Armenium," 124.
Ungulus, 215.
Ursula, St., relics of, 208.

Valerio Vicentino, crystals of, 116.
Venus, the Odescalchi, 115.
Venus-hair crystal, 301.
Verde di Prato, 257.
—— di Tarquinia, 207.
—— antique, marble, 257

ZMIRI.

Vermilion garnet, or Vermeille, 17
"Versicolor" glass, 337.
Vesuvian garnet, 17.
Vienna, agate vases at, 190.
Viper in amber, 308.
Viscera, human, agatized, 10.
Vishnu, the stone of, 255.
Vitriol, use of, 249, 314.
Voltaire in agate, 5.
Volterra alabaster, 223.
"Volucer," cup of L. Verus, 106.
——, supper of, 107.
Vopiscus, villa of, 128, 297.

Walsingham, the shrine at, 44.
Waterton Collection, the, 304, 322.
Water-sprites, to evoke, 122.
Wedgwood's pastes, 345.
Wellington, tomb of, 263.
Worms, used as a solvent for gems, 201, 355.

Xenocrates, 210.
Ximenes, 211.

Yellow, hair dyed to, 308.
Yu-stone, jade, 181.

Zachalias, 147.
Zenobia, zone of, 233.
Zircon, 161.
Zmiri, 172.

(G.) 2 c

LONDON:
PRINTED BY WILLIAM CLOWES AND SONS, STAMFORD STREET,
AND CHARING CROSS.